MW00437779

Contending Orders

Contending Orders

Legal Pluralism and the Rule of Law

GEOFFREY SWENSON

OXFORD
UNIVERSITY PRESS

Oxford University Press is a department of the University of Oxford. It furthers the University's objective of excellence in research, scholarship, and education by publishing worldwide. Oxford is a registered trade mark of Oxford University Press in the UK and certain other countries.

Published in the United States of America by Oxford University Press
198 Madison Avenue, New York, NY 10016, United States of America.

Library of Congress Cataloging-in-Publication Data
Names: Swenson, Geoffrey J. (Geoffrey Jon), 1981– author.
Title: Contending orders : legal pluralism and the rule of law / Geoffrey Swenson.
Description: New York, NY : Oxford University Press, 2022. |
Based on author's thesis (doctoral - University of Oxford, 2015) issued under
title: Addressing crises of order : judicial state-building in the wake
of conflict. | Includes bibliographical references and index.
Identifiers: LCCN 2022018856 (print) | LCCN 2022018857 (ebook) |
ISBN 9780197530429 (hardback) | ISBN 9780197530443 (epub) |
ISBN 9780197530436 | ISBN 9780197530450
Subjects: LCSH: Postwar reconstruction—Law and legislation—Afghanistan. |
Postwar reconstruction—Law and legislation—Timor-Leste. |
Nation-building—Afghanistan. | Nation-building—Timor-Leste. | Legal
polycentricity—Afghanistan. | Legal polycentricity—Timor-Leste. |
Afghanistan—Politics and government—2001– |
Timor-Leste—Politics and government—2002–
Classification: LCC KZ6787 .S94 2022 (print) |
LCC KZ6787 (ebook) | DDC 341.6/609581—dc23/eng/20220705
LC record available at https://lccn.loc.gov/2022018856
LC ebook record available at https://lccn.loc.gov/2022018857

DOI: 10.1093/oso/9780197530429.001.0001

1 3 5 7 9 8 6 4 2

Printed by Integrated Books International, United States of America

Contents

Figures

Tables

Acknowledgments

Since this book project began in 2012, it has benefited from the help of many individuals. I am immensely grateful to my DPhil supervisor at Oxford University, Richard Caplan, without whose assistance the project simply would not have been possible. Even before I was a student, he was immensely supportive: he responded to a cold inquiry from a random international development worker in Timor-Leste, with the constructive thoughtfulness that characterize all our discussions. My internal thesis examiner at Oxford, Nancy Bermeo, has been equally essential. She has offered invaluable guidance and support as this research transformed from a thesis into a book. I will always be grateful for the kindness and advice that both have generously offered over the years.

Also vital has been the support and guidance of my editor, Dave McBride, as well as the assistance of Holly Mitchell and the staff at Oxford University Press. The insights from the anonymous manuscript reviewers have also greatly improved the work. I am also grateful to OUP and the editors of *International Studies Review* for permission to draw on material from my article "Legal Pluralism in Theory and Practice" for Chapter 3. Likewise, my thanks to MIT Press and the editors of *International Security* for allowing me to use material from my article "Why U.S. Efforts to Promote the Rule of Law in Afghanistan Failed" for Chapter 7.

At Oxford, John Gledhill and Kalypso Nicolaïdis offered constructive insight that has dramatically improved the quality of my research, as has the input of Richard Youngs, my external thesis examiner. I am also thankful to Thomas Carothers, Erik G. Jensen, Silas Everett, and Russell K. Osgood for their support of this research from the start. The broader assistance of the Oxford community was vital. In particular, the Clarendon Fund and Pembroke College's Graduate Scholarship made it possible to pursue a DPhil at Oxford. During my fellowship in the International Development Department at the London School of Economics, Mayling Birney, J. P. Faquet, Teddy Brett, Pritish Behuria, Benjamin Chemouni, and Geoff Goodwin all helped refine and expand this research. Similarly, the project has benefited greatly from input from my colleagues at City, University of

London, especially Inderjeet Parmar, Amnon Aran, Sashi Sundaram, Sandy Hager, Koen Slootmaeckers, and Amin Samman.

Fieldwork in Afghanistan and Timor-Leste was invaluable for developing both my theoretical arguments and my understanding of key events. Travel grants from the Department of Politics and International Relations at Oxford University, Pembroke College, Oxford, and the LSE made that fieldwork possible.

Thanks is owed to Lillian Dang for her support of this research project, especially for sharing her invaluable insights on state and non-state justice in Afghanistan and Timor-Leste. Her logistical advice, along with that of the American Institute of Afghan Studies in Kabul, facilitated a level of access in Afghanistan far beyond what I could have reasonably expected. I also owe the staff of the Asia Foundation in Timor-Leste, particularly Susan Marx, Kerry Brogan, Diana Fernandez, Juliao Fatima, Todd Wassel, Gaspar H. da Silva, and Tony Ku, a debt of gratitude. In addition to logistical support, they offered keen and insightful analysis during my visits.

Several other friends and colleagues have greatly improved the project. My deep thanks to Meghan Campbell, Gabriel Rosenberg, Scott Shackelford, Frances Brown, Moritz Schmoll, William Partlet, and Kate Roll, as well as Francesca Ward, whose assistance was particularly invaluable. I also appreciate the support of Abi Williams, Lyal Sunga, Aaron Matta, Eamon Aloyo, Anca Iordache, and the rest of the staff at the Hague Institute for Global Justice. Last but not least, my deep gratitude goes to my parents, Karen and Lowell Swenson, and my sister, Carly Swenson. Any remaining errors and oversights are entirely my own.

Abbreviations

AATL	Assosiasaun Advogados Timor-Leste
ACC	Anti-Corruption Commission
AROLP	Afghanistan Rule of Law Project
CAVR	Commission for Reception, Truth and Reconciliation
CNRT	National Congress for Timorese Reconstruction
CNRT	National Council of Timorese Resistance
DFID	Department for International Development
DOD	Department of Defense
Fretilin	Revolutionary Front for an Independent East Timor
ICPC	Interim Criminal Procedure Code
IDLO	International Development Law Organization
INL	Bureau of International Narcotics and Law Enforcement Affairs
JHRA	Justice and Human Rights in Afghanistan
JSMP	Justice Sector Monitoring Programme
JSP	Strengthening the Justice System Program
JSSP	Justice Sector Support Program
LTC	Legal Training Centre
MSD	Management Sciences for Development
NATO	North Atlantic Treaty Organization
NCC	National Consultation Council
OECD	Organization for Economic Co-operation and Development
PDHJ	Provedor for Human Rights and Justice
RDTL	Democratic Republic of Timor-Leste
ROLFF-A	Rule of Law Field Force—Afghanistan
RLS-Formal	Rule of Law Stabilization Program—Formal Component
RLS-Informal	Rule of Law Stabilization Program—Informal Component
SCIU	Special Crimes Investigation Unit
SIGAR	Special Inspector General for Afghanistan Reconstruction
SJSA	Strengthening the Justice System of Afghanistan
SNTV	single nontransferable vote
UN	United Nations
UNAMA	United Nations Assistance Mission in Afghanistan
UNDP	United Nations Development Program

UNMISET	UN Mission in Support of East Timor
UNMIT	United Nations Integrated Mission in Timor-Leste
UNTAET	United Nations Transitional Authority in East Timor
UNTL	National University of Timor-Leste
USAID	United States Agency for International Development

PART ONE
LAW, ORDER, AND CONFLICT

This book critically examines how order is constructed, with a focus on democratic state-building and the rule of law promotion efforts in legally pluralist settings after conflict. It argues that the rule of law needs to be viewed holistically, and that engagement with non-state justice is a vital and understudied element in determining the success or failure of both domestic and international judicial state-building efforts. Beyond recognizing the importance of legal pluralism, it argues that the various ways in which state and non-state judicial authorities interact, and the consequences of those actions, need to be better understood. It has three main parts. Part One consists of the first three chapters. Chapter 1 introduces and defines the concepts of judicial state-building after conflict. Chapter 2 examines the rule of law and how it relates to legal pluralism. Chapter 3 outlines a new typology for understanding legal pluralism and the strategies employed to engage it.

1

Understanding the Challenge

Introduction

In late 2001 Afghanistan's future appeared brighter than it had in decades. Ever since the Communist coup in 1978, the country had been engulfed in endemic conflict. When the Communist regime finally collapsed in 1992, the country descended into violence between rival warlords. Eventually, this chaotic situation gave rise to the Taliban, who promised a regime of law and order premised on a very harsh interpretation of Islam. They largely succeeded. Success, however, carried a steep price. Afghans grew increasingly tired of the Taliban's brutality. The Taliban also provided safe harbor to Al-Qaeda terrorists, who planned and executed the September 11, 2001 attacks on the United States.

Shortly thereafter, international forces led by the United States began actively supporting the resistance. The Taliban's Islamic Emirate of Afghanistan quickly wilted away and was replaced by the Islamic Republic of Afghanistan. A multiethnic state under the leadership of President Hamid Karzai, it initially enjoyed robust domestic and international support. The new regime promised better governance for the Afghan people while also pledging to be a responsible global stakeholder. Yet that opportunity was squandered. The new government prioritized enriching itself and retaining power over building a legitimate democratic state capable of supporting the rule of law or even just simply trying to improve the lives of ordinary Afghans.

The legitimacy of the Islamic Republic of Afghanistan was soon contested. Since the founding of Afghanistan in 1747, all reasonably successful Afghan political and legal orders relied heavily on religious and tribal non-state judicial authorities who had long underpinned local order. In contrast, the new regime and its international backers were apathetic and sometimes overtly hostile toward them. As one senior USAID official explained to SIGAR staff, "We dismissed the traditional justice system because we thought it didn't have any relevance for what we wanted to see in today's Afghanistan" (SIGAR

Contending Orders. Geoffrey Swenson, Oxford University Press. © Oxford University Press 2022.
DOI: 10.1093/oso/9780197530429.003.0001

2018: 157). State courts cared far more about exacting money than seeking justice. Citizens still overwhelmingly preferred local shuras and jirgas to resolve disputes. By 2006 the Taliban had returned at the helm of an insurgency fueled by government corruption and lawlessness and underpinned by a parallel justice system that promised a legitimate and effective, if harsh, version of law and order. Taliban justice also capitalized on tribal structures and religious beliefs to build legitimacy.

Upon assuming office in January 2009, U.S. President Obama demanded a dramatic policy shift in Afghanistan. He sought to halt the steadily deteriorating security situation, bolster the state's failing legitimacy, and reverse the Taliban's striking resurgence. Promoting the rule of law, backed with military force, became a hallmark of counterinsurgency in the so-called Afghan surge. The official policy declared: "Justice and rule of law programs will focus on creating predictable and fair dispute resolution mechanisms to eliminate the vacuum that the Taliban have exploited" (U.S. Mission Afghanistan 2010: 9). Immense sums were spent to bolster the rule of law and Afghan state courts. However, these efforts proved futile, and the Taliban's insurgent justice system continued to gain ground. When the surge concluded in 2014, it was widely recognized as a failure. In the subsequent years, the prospect of a democratic state bound by the rule of law grew ever more distant. Ultimately, in 2021, President Joseph Biden determined that the U.S. mission in Afghanistan had reached its end. He announced the unilateral withdrawal of U.S. forces prior to the twentieth anniversary of the September 11 terrorist attacks. The Afghan regime established in their wake would not last even that long. The Taliban made rapid gains once the U.S. troop withdrawal began in earnest in July 2021. On August 15 the Taliban retook the capital with no organized resistance. President Ashraf Ghani fled. The Islamic Republic of Afghanistan had evaporated, and the Islamic Emirate of Afghanistan was reborn.

International policy makers made many mistakes. One key mistake was a failure to understand the consequences of legal pluralism when seeking to advance the rule of law in Afghanistan. There never was a justice vacuum. Local jirgas and shuras underpinned order in much of the country. At the same time, Taliban justice successfully asserted itself against the state in many places; it did not merely fill a void. Moreover, like the previous George W. Bush administration, Obama's team continued and dramatically increased funding to the state justice system, even though previous expenditures had achieved little. Despite billions in spending, decades of experience in post-conflict state-building, and numerous policy reviews, U.S. policy makers, and the international community more broadly, still failed to understand legal pluralism and its consequences for state-building. The Afghan state and its international backers faced deep but largely unacknowledged skepticism from traditional tribal-based authorities and outright combat with the Taliban justice system.

While Obama was proclaiming a new dawn in Afghanistan, Timor-Leste, another post-conflict state, faced its own stark challenges. After four hundred years of Portuguese colonial rule, the country endured a brutal occupation by Indonesia from 1975 to 1999 during which an estimated one-fifth of the population perished. After an independence referendum in 1999, another wave of violence from Indonesian-backed militias lead to vast death and destruction. International peacekeepers were required to stabilize the country. East Timor was placed under U.N. trusteeship until it gained independence in 2002. Like Afghanistan, the establishment of a new democratic regime in Timor-Leste was greeted with genuine optimism both domestically and internationally.

The new Timorese state sought to blend the independence movement's commitment to building a modern democratic state bound by the rule of law with institutions that respected and drew strength from the non-state dispute-resolution forums that remained the dominant form of justice. This was no easy task. The political alliances that proved so essential for the independence movement quickly frayed once that goal had been achieved. Tensions soon escalated between the leaders of the two main political groups led by Xanana Gusmão, who had led the CNRT, the umbrella organization of groups opposed to Indonesian rule, and Mari Alkatiri, the leader of Fretilin, the largest political party. Despite successful democratic elections and good-faith efforts to build a new, independent judiciary, violence broke out shortly after the departure of United Nations' peacekeepers. These events were dubbed "the 2006 crisis" or simply "the crisis." The crisis saw widespread rioting, violence, and internal displacement. It led to the collapse of Prime Minister Alkatiri's government and a return of U.N. forces to restore order. Even more daunting, this violence occurred between Timorese factions that had been partners during the independence struggle, raising the prospect of prolonged civil strife. Civil conflict was, however, avoided. After the democratically elected government collapsed, new elections in 2007 resulted in Gusmão's becoming prime minister. Nevertheless, the political situation remained tense, and the prospect of renewed violence was very real.

In 2008 the country was rocked by assassination attempts aimed at the president and prime minister. Political disorder prolifierated as law and order appeared to evaporate. Timor-Leste risked being labeled a failed state. But the Timorese state decidedly did not fail. Unlike their counterparts in Afghanistan, domestic policy makers redoubled their efforts to build a legitimate, democratic state and a state legal order, and to work constructively with non-state justice authorities. Timorese political leaders accepted the results of free and fair democratic elections. They recognized the symbolic and practical importance of regulating long-standing customary justice mechanisms in a way that was both legitimate and constructive through local democratic elections and an intuitive jurisdictional division between state courts and local village councils known as *sucos*. Non-state justice authorities largely embraced the ideals of the rule of law and democratic governance. Tensions remain, but since independence and especially after the 2006 crisis, the Timorese state has established a fundamentally cooperative, rather than competitive, relationship with the non-state justice sector.

International policy makers first prioritized supporting the physical infrastructure and human resources necessary to underpin the kind of modern justice system that had long been envisioned by Timorese leaders across the political spectrum. The 2006 crisis underscored the need for more constructive engagement between the state and non-state actors. International assistance helped bridge the gap. Both domestic and international efforts benefited from engaging with non-state justice on its own terms, rather than merely as an instrument to bolster the government or defeat insurgents. As a result, Timor-Leste not only stepped back from the abyss of civil conflict but made substantial progress in building a cooperative relationship between state and non-state justice alongside meaningful progress toward democracy and the rule of law more generally. While the state has certainly faced problems, Timor-Leste now ranks as a success story because state officials remained committed to translating rule of law ideals into laws and institutions. By 2012 the United Nations could proudly proclaim: "Today, Timor-Leste [is] . . . a place of peace, democracy, celebration and optimism, a place where new infrastructure, public investment and services are growing" (United Nations Integrated Mission in Timor-Leste 2012).

In their own ways, both Timor-Leste and Afghanistan reflect a central truth: state-building constitutes a major domestic and international endeavor with profound transnational security implications (Ciorciari 2021).[1]

[1] The record of international state-building efforts has been at best uneven and all too often profoundly discouraging (Jarstad and Sisk 2008; Paris and Sisk 2009a, b; Grimm and Leininger 2012).

As state-building seeks to establish governance institutions that can maintain law and order, establishing a viable state justice sector is essential (Paris 2004: 205–206). More broadly, the legal system embodies the state's authority and forms the foundation of its claims to sovereignty through its ability, or at least claim, to enforce a monopoly on the exercise of legitimate violence (Weber 1978). The legal system is thus vital to the state's ability to project state-sanctioned law over its territory and to secure power both domestically and internationally. While state-building can take many forms, this book focuses on judicial state-building.[2]

Robust legal pluralism was a defining feature of post-conflict Afghanistan and Timor-Leste. They are not alone. Indeed, it makes little sense to discuss promoting the rule of law without an appreciation of the larger justice ecosystem. Legal pluralism, where "two or more legal systems coexist in the same social field," can be found everywhere (Merry 1988: 870).[3] In developed countries, the state legal system's legitimacy and ability to resolve legal claims are taken for granted. But in many developing countries, alternative justice systems with independent authority—based on custom, tradition, religion, and a plethora of other sources—thrive. In fact, they handle the vast majority of disputes (Kyed 2011: 5). Non-state justice is particularly vital in conflict and post-conflict settings that have weak state institutions and contested governing authority.[4] In these situations, citizens do not or cannot

Some scholars go even further and argue that the international state-building enterprise itself is inherently illegitimate (Bain 2006; Chandler 2006) and counterproductive (Jahn 2007). While these criticisms deserve consideration, as Paris notes "for all the shortcomings of liberal peacebuilding—and there have been many—most host countries would probably be much worse off if not for the assistance they received" and forsaking state-building entirely would "abandon[] tens of millions of people to lawlessness, predation, and fear" (Paris 2009: 108). Regardless of how one ultimately conceptualizes state-building, the practice is here to stay for the foreseeable future. Thus, deepening scholarly understanding of what makes state-building more or less successful in environments featuring robust legal pluralism is certainly worthwhile.

[2] Judicial state-building refers to state-building activities focused on the legal system and, more broadly, the rule of law. Thus, judicial state-building is related to but distinct from state-building efforts seeking to establish, for instance, a robust military, mass education, or health care systems.

[3] What exactly constitutes law and legal pluralism is inevitably ambiguous (Griffiths 1986; Woodman 1998; Tamanaha 2021), but the idea of "folk law" is quite helpful. Tamanaha argues that "law is a 'folk concept"'—that is, law is what people within social groups believe it is in a given set of circumstances (Tamanaha 2008: 396). Thus, law outside of the state can take a variety of forms, including international law, customary law, religious law, natural law, *lex mercatoria*, and commercial arbitration (ibid.). My research, however, focuses on non-state justice systems that adjudicate criminal and civil matters generally handled by state courts in higher-capacity states. This non-essentialist definition of law is sufficiently expansive, but it avoids the pitfalls of maximalist approaches whereby basically all forms of social control are defined as law.

[4] The rare exceptions tend to be cases of state-versus-state conflict, such as Germany and Japan after World War II, where the losing state has been thoroughly defeated but still retains a strong preexisting legacy of a legitimate, high-capacity state (Fukuyama 2014: 80).

necessarily rely on the state justice system. Non-state alternatives are commonplace. They are frequently rooted in religion, custom, tribal identity or other normative structures. The authority to regulate people's conduct can be subject to fierce, even violent competition, and non-state justice may enjoy greater authority and legitimacy than the state justice (Migdal 1988; Swenson 2018a). Fostering the rule of law always faces serious obstacles, but settings where state capacity is limited and non-state justice authorities are autonomous, powerful, and respected can be particularly daunting.[5]

Non-state justice authorities can frustrate the state-building process or in extreme cases even present an existential challenge to the regime itself. State-building efforts, whether domestic, international, or a combination thereof, involve immense amounts of capital, time, and often lives. Misunderstandings about non-state justice and how domestic and international state builders interact with it are commonplace, dangerous, and repetitive. Nevertheless, the relationship between state and non-state justice remains insufficiently researched and decidedly undertheorized, especially in post-conflict settings.

This blind spot has led to limited scholarly understanding and, worse, suboptimal policymaking. This book illustrates how judicial state-building strategies could be tailored to better promote the rule of law in legally pluralistic societies after conflict. My argument draws on two case studies of major recent judicial state-building efforts in contexts of competitive legal pluralism with decidedly different results, Timor-Leste and Afghanistan. These cases can be used to begin to isolate factors that contribute to the success and failure of establishing the rule of law in post-conflict settings with extensive legal pluralism.

This book also offers a new theory by which to understand better how state and non-state justice systems interact with each other more generally. Specifically, it articulates four new legal pluralism archetypes: (1) combative, (2) competitive, (3) cooperative, and (4) complementary. These archetypes can help explain the dynamics of legal pluralism in any setting. In *combative* legal pluralism, state and non-state systems are overtly hostile to one another, such as occurred in Afghanistan between the Islamic Republic of Afghanistan and Taliban justice. *Competitive* legal pluralism features significant, often

[5] A stronger state legal system does not necessarily produce better policy or judicial outcomes. At a structural level, however, only by the state system's reaching a workable arrangement with its non-state counterparts can the core facets of the rule of law be achieved, as these require a baseline level of consistency and uniformity in the application of justice.

deep tensions between state and non-state legal systems, but clashes rarely endanger the state's formal judicial authority because the non-state justice sector does not make a concerted effort to supplant that authority. This dynamic was present in places like Jordan under the Hashemite monarchy (Al Naber and Molle 2016) or Tunisia after the Arab Spring (Salim 2015). In a *cooperative* legal-pluralist environment, non-state justice authorities still retain significant autonomy and authority, as in contemporary Timor-Leste. However, state and non-state legal authorities are generally willing to work together toward shared goals. Finally, under *complementary* legal pluralism, non-state justice authorities operate under the umbrella of state authority and without substantial autonomy to reject state decisions. For example, in the United States, Native American tribal law operates with exemptions from certain state legal regulations but still under the overarching legal authority of the U.S. government.

Understanding legal pluralism can be daunting, but it is even more challenging to influence it. This book identifies and critically examines commonly used strategies in legally pluralistic environments. Moreover, it articulates the most promising approaches under different archetypes by drawing on extended case studies from two states with extensive legal pluralism, Afghanistan and Timor-Leste. Timor-Leste has moved from competitive to cooperative legal pluralism. Afghanistan deteriorated from competitive to combative legal pluralism before the Taliban achieved outright control of the state. These two cases showcase divergent institutions, strategies, and policy decisions and thus offer crucial insights into understanding post-conflict judicial state-building. Vibrant legal pluralism and the dynamics between state and non-state systems directly influence the results of international and domestic efforts to promote the rule of law after conflict. Judicial state-builders must accurately determine the legal-pluralism archetype and its programmatic implications, along with developing a credible strategy for transforming the current environment into a more constructive situation.[6] Of course, while archetypes can offer valuable guideposts, each approach will need to be cognizant of the local context. In other words, all situations of competitive legal pluralism are competitive, but the exact parameters and characteristics of that competition will vary.

[6] Despite the importance of strategies, policies, and programs related to state and non-state justice actors, the relationship is not strictly linear. Broader social, economic, and political factors also have important consequences.

This chapter explains the research and methodological approach. It starts with a research overview. My approach differs from standard academic treatments and clarifies some key concepts. Next, it explains the use of case studies and the collection and analysis of sources and data. Finally, it describes the book's overall structure and arguments.

Overview of Research

Establishing the rule of law is widely seen as foundational for the success of state-building efforts by scholars (Paris 2004: 206; Blair 2019) and policy makers ranging from former U.S. President George W. Bush (Bush 2002: 9) to U.N. Secretary-General Kofi Annan (Annan 2004: 3). In post-conflict states and many developing countries more generally, however, non-state justice is essential for maintaining order and resolving most legal disputes. For most people in these states, non-state justice is the primary forum for justice. Thus, this book seeks to answer two questions: what are the challenges to the promotion of rule of law in legally pluralistic post-conflict settings, and what accounts for the success and failure of domestic and international state-builders to meet those challenges?

Although the answer is no doubt complex and to some extent context specific, I argue that first and foremost state officials must display a normative commitment to the rule of law in both the actions they take and the institutions they create and sustain. Political and judicial leaders must take the law seriously. They must also support the development of institutions or "rules of the game in a society" that promote accountability, including democratic governance underpinned by free and credible elections, the creation of viable political parties, and an independent judiciary (North 1990: 3). The rule of law entails both top-down elements that reflect the importance of leadership and bottom-up elements that illustrate the vital role of popular demands for representation and accountability. Second, absent a successful effort to repress non-state justice, the state must be willing to meaningfully engage with at least some powerful non-state actors. This invariably involves a degree of compromise. Non-state-justice actors take many forms, but they are often rooted in religion, custom, geographic difference, or tribal affiliation. In extreme cases, that authority can be based largely on force, as exemplified by warlords in Afghanistan.

Non-state judicial actors tend to derive their authority from sources beyond the state. State officials may seek to ratify, regulate, or destroy that non-state authority. Or they may simply ignore them. State officials can also play a crucial mediating role among non-state actors and the international community owing to their more direct, influential relationship with non-state actors. Even when external actors would like to work directly with non-state-justice authorities, state officials retain a great deal of influence over whether and how foreign assistance is distributed (Swenson 2018b). East Timorese state officials were imperfect but largely effective in mediating between the international community and local-level figures, a relationship that contributed to the transformation of an environment of competitive legal pluralism into one of cooperative legal pluralism. In contrast, although the Karzai-led regime in Afghanistan pledged to establish a democratic state bound by the rule of law, it was not actually committed to those ideals, and neither state officials nor their international backers ever seriously sought to engage with non-state-justice actors, especially powerful tribal authorities, who were long-standing pillars of legitimacy. In contrast, the Islamic Republic and its international backers were far too willing to work with warlords who promised to deliver security, despite their general lack of legitimacy and obvious disinterest in democracy and rule of law.

International actors are far from powerless. They can offer incentives (or disincentives) for both state and non-state-justice actors to encourage certain behaviors or policies. Conversely, non-state judicial actors must be receptive to state and international judicial state-building efforts. All three actors are essential (though not sufficient) for success. Non-state actors may be more or less open to working with state and international actors. Non-state legal authorities' willingness to engage is particularly vital when they are strong enough to successfully resist state- or internationally led initiatives, as is often the case in post-conflict and conflict-prone settings. As Migdal explains, "The major struggles in many societies, especially those with fairly new states, are over who has the right and ability to make the countless rules that guide people's social behavior" (1988: 31). State institutions and policy decisions matter immensely. At the same time, whether non-state justice sector actors choose to constructively engage with international judicial state-building initiatives has a significant impact on their outcomes.[7]

[7] While this book explores legal pluralism theoretically and empirically, it does not attempt to provide a detailed history or determine the quality of state versus non-state legal systems. Non-state justice systems may provide just or unjust outcomes in particular cases and can potentially foster either

Conceptualizing the Rule of Law and State versus Non-State Justice

While the rule of law, non-state justice, and the state justice system are discussed more extensively in Chapter 2, it is important to clarify at the outset what is meant by these terms. Given the immense challenges they face, it is most appropriate to assess post-conflict states under the criteria of a "thin" rather than a "thick" conceptualization of the rule of law. At minimum, a thin concept, the rule of law requires that "law must be set forth in advance (be prospective), be made public, be general, be clear, be stable and certain, and be applied to everyone according to its terms" (Tamanaha 2007: 3). Establishing a minimalist understanding of the rule of law after conflict is difficult and slow. It demands laws, judicial institutions, and legal personnel alongside the ability to enforce state regulations nationwide and an ideological commitment to the concept of the rule of law (Stromseth, Wippman, and Brooks 2006: 76). Thicker conceptualizations of the rule of law require meeting these conditions alongside certain fairness-oriented institutional, economic, cultural, and political requirements (West 2003).

The state justice system refers to the institutionalized mechanisms for implementing and enforcing state-sanctioned law. It consists of courts and court personnel, most notably judges, prosecutors, defense and other attorneys as well as the state entities set up to engage directly with the judiciary, such as a ministry of justice (MOJ) or a prosecutor-general office. This definition is broad but not boundless. It excludes actors with important roles in the system, such as legislatures that draft the law or correctional systems that oversee individuals convicted of violating it.

Non-state justice refers to those legal systems that credibly structure and potentially sanction behaviors that have significant autonomy from state law (though state officials may sometimes acquiesce in its use).[8] These rules and regulations are generally based on custom, tradition, religion, family lineage, and other powers not sanctioned by law (such as criminal enterprises

social stability or social instability at the local level. Likewise, it is wrong to assume that outcomes are more or less just simply because they stem from the state justice system or that the non-state system is more or less representative of the population's preferences.

[8] The use of the word "non-state" reflects a conscious decision to avoid terms such as "informal," "traditional," or "customary," all of which reflect empirical and often normative claims, unless they are demonstrably accurate for a given non-state legal system.

or powerful commercial interests). They may have equal or even greater force than state law. Rule violations may lead to severe sanctions and even death.

Understanding the Relationship between State and Non-State Justice Systems

This book proposes new archetypes to help us better understand the relationship between state and non-state judicial actors as discussed in greater depth in Chapter 3. Of the four main archetypes, three are particularly relevant to post-conflict settings: combative, competitive, and cooperative legal pluralism. Complementary legal pluralism is more likely to be found in places where peace has been achieved. In situations defined by combative legal pluralism, the state and non-state justice systems are overtly hostile. Unsurprisingly, combative legal pluralism is commonly found in countries facing an active insurgency or separatist movement. In many instances, non-state justice forms a cornerstone of challenges to the state's authority (Mampilly 2011).

Competitive legal pluralism is the default setting in most post-conflict environments because these tend to have relatively weak state institutions and contested governing authority. Competitive legal pluralism features significant, often deep, tensions between state and non-state legal systems, but these tensions rarely endanger the state's overarching formal legal supremacy. Competitive legal systems are frequently rooted in religious beliefs or shared culture, custom, or heritage that does not necessarily reflects the state's official values. In a cooperative legal-pluralist environment, non-state judicial authorities retain autonomy and authority but are usually open to working together toward shared goals. Major clashes between state and non-state actors are infrequent and tend to reflect social issues, such as women's rights, rather than existential issues of state judicial power. Cooperative legal pluralism flourishes where progress is being made toward consolidating legitimate state authority. Non-state justice systems retain real autonomy but nevertheless largely support the state. Under complementary legal pluralism, state authority does not face a meaningful challenge from non-state judicial actors, which is a rarity in post-conflict states.

The overarching legal-pluralism archetype influences how post-conflict judicial state-building enterprises should be understood and pursued, because it can substantially constrain or bolster domestic and international initiatives. It can also offer insight into how to structure particular laws, policies, and programs. At the same time, these archetypes are neither set in stone nor determinative. Both cases started in situations of competitive legal pluralism. Timor-Leste shifted to cooperative legal pluralism over time, while Afghanistan descended into combative legal pluralism. Indeed, it was the Taliban's rival system of justice rather than that of the duly recognized state that ultimately triumphed. Effective engagement with the non-state justice sector requires thinking critically about how to deal with current realities while simultaneously seeking to transform the environment into one more favorable to judicial state-building.

The Relationship between Democracy and the Rule of Law

Democratic rule does not inevitably produce the rule of law, but it is a functional prerequisite. The rule of law is not apolitical or technocratic in origin. The legal order is inexorably connected to a state's political institutions (Weingast 1997; Acemoglu and Robinson 2012). The case-study chapters examine the state legal system and non-state alternatives as well as how they relate to the larger governance structures in Afghanistan and Timor-Leste that supported or failed to support democratic governance and progress toward the rule of law.

While the rule of law is deeply linked with democratic government, the concepts are distinct. In theory, the "rule of law may exist without democratic forms of political will formation," but on "empirical grounds" the rule of law has been interconnected with democratic government (Habermas 1995: 12).[9] As O'Donnell states:

> Without a vigorous rule of law, defended by an independent judiciary, rights are not safe and the equality and dignity of all citizens are at risk. Only under a democratic rule of law will the various agencies of electoral, societal, and

[9] Singapore is occasionally cited as an expectation that offers both non–fully democratic governance and the rule of law, a combination Rajah dubs "authoritarian rule of law" (Rajah 2012). This view, however, conflates state performance with the rule of law. While the legal order is largely predictable, opposition is effectively sidelined; furthermore, "there is neither a free press nor an autonomous civil society, and divergence of political views is not readily tolerated" (Mutalib 2000: 316). The law does not apply equally to everyone. A legitimate legal order protects economic rights and delivers high-quality governance, but its core purpose is perpetuating the existing political regime, which is characterized by "authoritarian dominance" (Slater 2012: 23).

> horizontal accountability function effectively, without obstruction and intimidation from powerful state actors. (O'Donnell 2004: 32)

Assessing the development of the rule of law requires examining judicial institutions alongside the major political institutions that must sustain the rule of law over time. Courts certainly play an essential role in constitutional democracy. They punish crimes, resolve private disputes, and help maintain order. They can be a bulwark again excessive executive power, contribute to a culture where state officials are bound by law and legal norms, and safeguard individual rights. The structure of the executive branch, the political party system, the election system, and the broader political and legal culture are also vital.[10] For instance, excessive executive authority can overwhelm even an independent judiciary with relative ease. Elections can help build rule-of-law culture through both normalizing the peaceful transfer of power and facilitating broader political participation.

Non-state justice serves as a cross-cutting factor across all these areas as a major forum for dispute resolution and a potential judicial state-building spoiler. This influence cuts both ways. The state's institutions, level of legitimacy, and stance toward non-state justice influences the operations of non-state justice as well as the posture of non-state-justice actors toward the state. These can be straightforward such as whether state officials want a collaborative relationship with non-state actors. These can also take surprising forms. For instance, the Timor-Leste case study suggests that one of the state's most effective tools for engagement is helping to broker free, fair, and competitive democratic elections for local justice authorities. Key local elections for non-state actors proved vital for establishing a cooperative rather than a competitive relationship because it established clear incentives to work with the state.

Legitimacy and the Rule of Law

The existence of the rule of law presupposes a legitimate state legal order. History suggests that the rule of law takes roughly fifty years to develop

[10] The importance of these factors has been widely noted: on executive authority, see Stepan and Skach (1993), Linz (1994), Tsebelis (2002); for electoral systems, Sartori (1997), Reynolds (2002), and Norris (2004); regarding political parties, Lipset (2000), Diamond and Gunther (2001), and Carothers (2006a); and a rule-of-law culture, Harrison and Huntington (2000), Stromseth, Wippman, and Brooks (2006), Licht, Goldschmidt et al. (2007), and Tamanaha (2011a).

under even highly favorable circumstances (North, Wallis, and Weingast 2009: 27). A legitimate legal order, then, is necessary but not sufficient for establishing a state bound by the rule of law. Legitimization comes in a wide variety of forms (Beetham 2013). People view the political and legal order as legitimate and authoritative for a wide variety of reasons, such as military success or religion, even if the state's democratic credentials are questionable or nonexistent (Mann 2012). Thus, the key issue is not whether the rule of law has been established in the years following conflict, but (1) whether a legitimate state legal order exists and (2) whether there has been progress toward consolidating the rule of law. This should not be confused with claims regarding whether the rule of law exists. Even in the successful case study, Timor-Leste did not meet even a minimal rule-of-law ideal. Nevertheless, real progress had been made as state officials, with support from both non-state justice authorities and international actors, constructed a legitimate political and legal order capable of supporting continued movement toward the rule of law.

The Study of Post-Conflict Judicial State-Building and Legal Pluralism

This section puts forward two main contentions. Existing scholarly explanations are inadequate for understanding the legal, political, and cultural backdrops against which judicial state-building occurs, and for understanding how the policies and strategies used by domestic and international actors influence outcomes. An alternative understanding attuned to legal pluralism's structure and importance would provide more robust analysis, but even when scholars draw on the idea of legal pluralism and judicial state-building, the analysis is incomplete. This book addresses these issues.

The non-state system is usually seen as either an obstacle to progress that must be addressed (Farran 2006) or as something that is effective and efficient, and reflects a community's true preferences (Harper 2011a). For instance, a major U.N. report, "Informal Justice Systems," defines non-state justice systems as involving "a neutral third party not part of the judiciary" while also noting that "custom-based systems appear to have the advantages of sustainability and legitimacy" (UNDP, U.N. Women et al. 2012: 8). Non-state justice is often idealized as a legitimate, advantageous, and cost-effective alternative order where "law is only law to the extent that the people who follow it, voluntarily or otherwise, consider it to have the status of law" (Harper 2011a: 17). In both instances, the binary between the state and non-state justice sector oversimplifies complex and contested relationships. While it makes sense to talk about state and non-state justice systems, they frequently intermingle in practice.

When determining how and to what extent to engage with the justice sector as a whole, judicial state-building strategies should assess whether the legal-pluralism environment is combative, competitive, or cooperative. This is in contrast to the tendency to assume a generally static or at least largely independent relationship between the two systems.

The Failure of Commonly Cited Explanations to Explain the Divergent Outcomes

Afghanistan and Timor-Leste have generated no shortage of commentary. Yet, the many frequently cited explanatory factors, which include the level of international investment, the states' divergent histories and cultures, their propensity for conflict, and their differing levels of state capacity, cannot fully explain the divergent outcomes in the two countries. First, the level of external support alone cannot explain the different results. While the exact figures are hard to ascertain, it has been recognized that the monetary costs involved in each intervention have been enormous (Butler 2012; Greentree 2013). It is not simply that one state had significantly more assistance than the other and was therefore able to succeed. While insufficient assistance may preclude success, more external assistance alone certainly does not guarantee it and may even be counterproductive (Barnett and Zürcher 2009; Maley 2018).

Second, the results in Afghanistan and Timor-Leste were by no means preordained, either by history or culture or by an inclination toward conflict. While it is now regarded as a success by many observers, in the beginning state-building efforts in Timor faced fierce, sustained criticism (Chopra 2002; Richmond and Franks 2008). The transition to local governance "was punctuated by initially sporadic, and then intense, episodes of violence" (Gledhill 2012: 48). The events surrounding the 2006 crisis cast grave doubt on the state-building enterprise (Richmond and Franks 2008: 48; Scambary 2009) and many feared the new state was destined to fail (Cotton 2007). Although there is often a fatalism about Afghanistan, state-building efforts started with great optimism. As the longtime scholars of Afghanistan Ahmed Rashid and Thomas Barfield have persuasively argued, a meaningful opportunity did exist for democratic state-building efforts to be successful after the Taliban regime's colipase (Rashid 2008; Barfield 2010:300–310). More broadly, for all their faults, the first Taliban regime and the earlier experience under the monarchy demonstrated that it was possible to establish a legitimate state legal order.

Finally, differences in existing governance structures alone do not explain the outcomes. Prior to the Soviet invasion in 1979, the Afghan state enjoyed greater capacity to maintain order and provide public goods (Rubin 2002). After 1975 East Timor had a strong governance structure under Indonesian rule, but the state was seen as illegitimate and a colonial imposition rather than an organic governance structure (Robinson 2009:21–91). Furthermore, existing state infrastructure was almost entirely destroyed in the violence surrounding the 1999 independence referendum (Downer 2000:8–9). Both states had extremely limited capacity when conflict ceased.

The Insufficiency of Current Understanding of Legal Pluralism's Implications for State-Building

Beyond the specific cases, the literature on this subject is largely historical, theoretical, or focused on how particular non-state justice systems operated. The recognition of competition between the state and non-state justice systems dates back to the seminal work of the legal historians Pollock and Maitland, who noted the inevitable tension generated by multiple legal systems and the state's desire to expand its judicial power at the expense of customary systems in England (Pollock and Maitland 1923: 184–186). Drawing on their work, Francis Fukuyama contends that competition between legal systems drove the formation of the modern state (Fukuyama 2011:245–275). Legal scholars have noted that when legal pluralism is prevalent, "these coexisting sources of legal authority, these coexisting sources of normative ordering are poised to clash particularly when their underlying norms and processes are inconsistent" (Tamanaha 2008: 400). Anthropologists have long been attuned to the existence of multiple justice systems, and to the fact that non-state justice can be the dominant legal order and operate very differently from state law (Malinowski 1985 [1926]).

In Timor-Leste, scholars have recognized the existence and importance of non-state justice as a dispute-resolution forum (McWilliam 2005, Nixon 2012). Regarding Afghanistan, Barfield has argued that clashes between legal systems generated conflict as "for over a century, successive national governments in Kabul have sought to impose centralized code-based judicial institutions upon local communities that have historically had their own informal institutions for regulating behavior and resolving problems" (Barfield 2008: 349). While Barfield and others have useful insights into the function of, and competition between, differing legal systems, relatively little work examines international judicial state-building initiatives and how they can influence, or have influenced, the competition between legal systems. Moreover, as discussed in greater detail in Chapter 3, the existing research does not account for different types of relationships between state and non-state justice and the fluidity between different legal-pluralism archetypes.

Scholars rarely examine the impact of legal pluralism on contemporary judicial state-building endeavors.[11] Fukuyama's work, for example, does not discuss how international state-building influences the competition between legal systems (Fukuyama 2004). Scholars rightly note that international state-building efforts occur in the context of "hybrid institutions" (Mac Ginty 2008), where external actors can significantly influence the behavior and incentives of local elites and other stakeholders (De Waal 2015), and that legal pluralism is inherently fluid and responsive (von Benda-Beckmann and von Benda-Beckmann 2006). Still, there has been little empirical investigation into and even less discussion in terms of theory of how domestic and international state-building efforts influence the overarching relationship between state and non-state justice, as well as how those interactions influence progress toward establishing the rule of law. The emphasis remains on the relationship of legal pluralism to development initiatives more generally (Tamanaha, Sage et al. 2012) or on attempting to understand the prevalence of legal pluralism (Glenn 2011). Moreover, little systematic work examines its implications for international and domestic judicial state-building efforts, particularly scholarship. Nonetheless, there is a growing literature that examines related issues such as the history and expansion of legal pluralism

[11] While this research focuses on normal justice, it is important to note that transitional justice frequently becomes a pressing concern after conflict or a regime change. As such, it has spawned a vast literature in general (Vinjamuri and Snyder 2015) and in relation to judicial state-building (Sriram 2007; Tricia, Leigh et al. 2010) and non-state justice (Clark 2010; Kochanski 2020). The "extraordinary measures" of transitional justice, however, are analytically distinct from efforts focused on "normal" criminal justice (McEvoy 2007: 422).

and its role in development (Benton 1994; Grenfell 2006; Isser 2011b; Tamanaha 2011l Berman 2012) as well as practitioner literature addressing how best to engage non-state justice systems to achieve ends other than judicial state-building (DFID 2004; Barfield, Nojumi et al. 2006; Wojkowska 2006; Harper 2011b; UNDP, U.N. Women et al. 2012). While recognizing these important contributions, this research takes a different approach by focusing on the role of legal pluralism in post-conflict judicial state-building, both domestically and internationally, and the use of process tracing.

Case Studies

This book presents case studies "as an in-depth study of a single unit (a relatively bounded phenomenon) . . . to elucidate the features of a larger class of similar phenomena," in this instance post-conflict judicial state-building in legally pluralist environments (Gerring 2004: 341).[12] It makes a structured comparison of approaches to the state and non-state justice sectors after conflict using process tracing (George and Bennett 2005:114–120). More specifically, this book explores post-conflict judicial state-building endeavors during periods of major international involvement, approached through a structured comparison. It examines how domestic and international state-builders sought to establish the rule of law through engagement with both the state and non-state sector and the consequences of those actions. Through a mixture of "theory development" (Timor-Leste) and "theory testing" (Afghanistan) (George and Bennett 2005:114–120), it seeks to explain, at least in part, what accounts for the success or failure of judicial state-building efforts where non-state judicial actors retain substantial authority and autonomy.

Process tracing examines the "links between possible causes and observed outcomes" to determine key causal mechanisms and the most plausible findings (Bennett 2010: 6). A case-study approach that uses process tracing

[12] For this research, a case-study approach offers compelling advantages over quantitative (large-*N*) approaches. First, a case study can capitalize on "a full variety of evidence—documents, artefacts, interviews, and observations—beyond what might be available in conventional historical study" (Yin 2009: 11). Case studies are better than large-*N* studies for rarer events that may nevertheless possess profound importance (Gerring 2007: 56). Post-conflict state-building initiatives against the backdrop of international intervention are relatively rare, but they tend to be in places marked by combative or competitive legal pluralism. Moreover, they are high-stakes enterprises involving staggering sums of money and effort (Call and Cousens 2008: 1–2).

can help illuminate judicial state-building processes and outcomes within as well as between cases.[13] It will highlight key developments and illuminate how the archetypes and strategies discussed in Chapter 3 can offer important insights into how relationships between state and non-state actors can change for the better or for the worse. Here it involves a detailed, close examination of clearly bounded domestic and international state-building initiatives in the legally pluralist environments of Afghanistan and Timor-Leste. For both theory-building and testing, extensive examinations offer advantages over abbreviated case studies that may not "provide sufficient detail for a rich and contextualized understanding of the case and processes, and they surely do not capture all the causal dynamic or even all the variables that a lengthier treatment might identify" (Call 2012:16–17).

Timor-Leste and Afghanistan are paradigmatic examples of post-conflict, competitive legally pluralistic settings where domestic and international policy choices have had major consequences. From the outset, policy makers themselves recognized that these cases would have broad significance and inform future endeavors (Chesterman 2002, Chopra 2002). Timor-Leste was explicitly envisioned as a model for post-conflict Libya, though that option was ultimately rejected (Doyle 2016: 26). Even before the Islamic Republic of Afghanistan disintegrated, policy makers were fixated on trying to learn lessons from Afghanistan (Hassan 2020).

Because any commitment to supporting the rule of law is contextual and depends on institutions that are based and built upon history and culture even if they are ostensibly new, it is essential to understand the foundations of legitimacy (Fukuyama 2011, 2014). The process of developing the rule of law varies dramatically by country (Acemoglu and Robinson 2012). The development of the rule of law and democracy takes time and can be understood only in a historical and cultural context (Cao 2016). An attempt to inaugurate a robust, legitimate state bound by the rule of law must take into account the established pillars of legitimacy in that society as well as why past attempts to build such a state may have failed. The case-study chapters for Afghanistan and Timor-Leste illustrate relevant history and culture and underscore that multiple plausible paths to advancing democratic governance and the rule

[13] Robust contemporary documentation and the ability to interview key actors makes process tracing feasible. It permits additional methodological rigor, helps generate new insights, and tests those insights' plausibility, as well as increasing the explanatory power of potential explanations while helping to eliminate competing accounts (Tansey 2007: 771). Qualitative methods, including case studies drawing on process tracing, "can add inferential leverage that is often lacking in quantitative analysis" (Collier 2011: 823).

of law exist. These paths, however, invariably reflect the history, institutions, culture, and structure of the earlier law and governance arrangements and non-state justice systems that existed prior to the conflict as well as those that emerged during it. These chapters will also serve to debunk certain key arguments regarding state-building in Afghanistan and Timor-Leste.

Other important similarities exist that further enhance the value of a comparison between these states. Afghanistan and Timor-Leste easily rank among the most ambitious judicial state-building efforts undertaken in recent memory. Both countries established new regimes and new legal systems in the early 2000s in contexts marked by competitive legal pluralism where the state justice sector needed to demonstrate its value, effectiveness, and legitimacy. The vast majority of disputes in both countries were settled outside the state justice system. Dispute resolution most commonly occurred at the community level through non-state mechanisms, including jirgas or shuras in Afghanistan and suco councils in Timor-Leste (Barfield, Nojumi et al. 2006; Everett 2009). In both states, the prevalence of the non-state justice sector, unsurprisingly, coincided with a history of limited state capacity and weak central rule (Murtazashvili 2016; Robinson 2009). Non-state justice systems faced few jurisdictional restrictions or meaningful attempts by the state to regulate their conduct, particularly outside of urban centers. Both states saw their infrastructures devastated by conflict and faced high levels of poverty (World Bank 2003; UNDP 2004). They enjoyed a period of stability, but the prospect of renewed violence remained as each country endured repeated bouts of upheaval and violence. In Timor-Leste, the 2006 crisis cast serious doubt on state-building endeavors and even the viability of the state itself. After the first Taliban regime was defeated, Afghanistan faced not only the prospect of renewed conflict, but also a rare opening to support democratic governance and a more inclusive order capable of supporting progress toward the rule of law. The country's history demonstrates that peace is possible. Prior to the 1978 Communist coup, Afghanistan enjoyed decades of domestic tranquility. After a period of prolonged violence, the first Taliban regime monopolized the legitimate use of force over most the country. After the 2001 Bonn Agreement, Afghanistan enjoyed relative peace before the Taliban insurrection became a full-scale insurgency in 2006.

In both cases, local and international state-builders invested heavily in the state justice sector, and similar initiatives were undertaken. International and domestic judicial state-building activities focused on (1) establishing or reforming institutions, (2) drafting laws and regulations (3) strengthening the legal profession and auxiliary institutions, and (4) increasing legal access and advocacy. These efforts were all led by the state, but usually in collaboration with international actors. Such interventions are commonplace in rule-of-law efforts (Carothers 1999: 168). Thus, Afghanistan and Timor-Leste are broadly representative.

The divergent shifts in the legal-pluralist environment were not inevitable. As the case-study chapters demonstrate, the decisions of domestic state officials and international actors really mattered. These differences offer important insights into understanding judicial state-building under different legal-pluralism archetypes. Thus, each case study comprises two chapters: the first examines domestic judicial state-building efforts in their historical context, while the second examines international judicial state-building initiatives.

Despite these shared circumstances, the two countries have radically diverged during the period examined. Timor-Leste successfully shifted to cooperative legal pluralism, while decisions in the former regime in Afghanistan helped trigger a slide to combative legal pluralism with the resurgent Taliban's justice system. Notwithstanding the upheaval surrounding the 2006 crisis, judicial state-builders in Timor-Leste made significant progress toward developing the rule of law by working to build an effective state justice sector under the auspices of a legitimate democratic state and collaborating with local non-state authorities. In contrast, the behavior of the post–Bonn Agreement regime helped trigger a slide from competitive legal pluralism to combative legal pluralism owing to state officials' lack of commitment to democratic governance and the rule of law and their failure to engage key non-state actors, who ultimately played an important part in the collapse of the regime itself.

These differences reflect both domestic and international decisions. In Timor-Leste, political elites supported a vision of a democratic state bound by the rule of law, as embodied in the independence movement (Robinson 2009). This vision was translated into a vibrant competitive multiparty democracy underpinned by a constitutional order. International engagement generally reinforced domestic trends promoting the rule of law, by supporting the development of the state justice system and constructive engagement with the non-state justice system. In Afghanistan, after the first Taliban regime fell, the new Islamic Republic of Afghanistan, headed by Hamid Karzai, enjoyed a promising opportunity to garner legitimacy as well as to advance democracy and the rule of law. Yet, upon acceding to power—with international backing—Karzai's regime quickly undermined democratic and legal accountability mechanisms. Likewise, the regime did not make a good-faith effort to work with the tribal and religious authorities that had long underpinned legitimate legal order and governance more generally. International actors did not seriously attempt to arrest these negative trends and occasionally exacerbated them.

International judicial assistance is designed, negotiated, and implemented in a particular place and time. Thus, despite broad similarities in approach, each judicial state-building situation is unique. Specific rule-of-law programs should be compared to better understand interactions with the non-state justice systems at a local level. How programs influence broader judicial state-building outcomes must also be examined. Aid to Timor-Leste from the United Nations, the United States, and Australia achieved generally positive results because their support reinforced sound domestic efforts to build the rule of law and meaningfully engaged the non-state justice system. Portuguese assistance, however, was a problematic exception. Aid from Portugal sought to institutionalize the Portuguese language throughout Timor-Leste's state institutions, and especially its legal institutions. It proved vital for establishing a modern judicial system. At the same time, this aid raised issues of accessibility and fairness. Nearly two decades after achieving independence in 2002, still only a tiny minority of Timor-Leste's population spoke Portuguese. While domestic and international policymakers supplied ample rhetoric about the importance of the rule of law in Afghanistan, there was never a credible strategy for actually promoting it. Massive investment produced no discernible progress toward the rule of law. Likewise, international engagement with the non-state justice sector achieved little because it was based almost exclusively on counterinsurgency priorities rather than promoting the rule of law. International actors helped enable a predatory and corrupt domestic regime, which in turn helped pave the way for the Taliban's eventual return to power.

Sources and Materials

The case studies draw on a mixture of primary and secondary sources relating to state-building, the rule of law, and legal pluralism generally and in Timor-Leste and Afghanistan specifically. My research uses primary sources produced by the states themselves, international organizations, donors, and nongovernmental organizations (NGOs). Reports produced by implementers, donors, and government agencies can illuminate their strategies, activities, and goals. They reflect assumptions regarding causal mechanisms as well as the link between their activities and desired change. These materials articulate clear views on the aspiration and justifications of key actors and institutions, which can then be critically examined. Domestic and international state-builders are quite forthright about their objectives (though not necessarily their motivations). This makes it possible to trace state-building efforts over time and examine how actors responded, or failed to respond, to unfolding events, as well as to critically inspect the ends pursued and means selected. Donors, states, and implementers often portray their activities in the best possible light, so their conclusions should not be accepted uncritically. Together these documents provide concrete empirical information, as funders tend to demand that implementing organizations provide clear data regarding program activities, outcomes, political context, and narrative framing, along with statements regarding overarching strategy and expected results.

Post-conflict settings raise significant research challenges (Roll and Swenson 2019). In 2014 and 2015 I interviewed in person over forty key stakeholders related to post-conflict judicial state-building and legal pluralism during my field research in Afghanistan and Timor-Leste. I did additional fieldwork in Timor-Leste in 2017 and 2019.[14] Interviews were semi-structured and combined purposive sampling with chain referral sampling.[15] Chain referral sampling seeks references for additional interviewees from initially identified respondents that might not be immediately apparent, as is frequently the case when dealing with NGOs and state or international bureaucracies (Farquharson 2005). The case studies also draw on insights from working on judicial state-building initiatives in both countries between 2008 and 2012. These cases possess clear temporal and spatial boundaries. The temporal boundaries result from the extent of large-scale international involvement after conflict, particularly the presence of significant military

[14] While all fieldwork involves risk (Swenson and Roll 2020), unfortunately the level of risk in Afghanistan made it infeasible to undertake independent in-country research there after 2014. Additional research interviews conducted since then took place electronically or in person but outside Afghanistan.

[15] Tansey argues this constitutes the "optimal" approach for case studies utilizing process tracing (2007: 771). First, interviewees, particularly in the initial stages, were chosen through a purposive approach that selects "respondents according to the position they have held and their known involvement in the political process," which in this case are the domestic and international judicial state-building efforts (Tansey 2007: 770).

forces.[16] The Timor-Leste case covers independence in May 2002 through to the exit of international forces in 2012. The Afghanistan case study covers the start of the post-Taliban transitional administration in late 2001 through to the election of Karzai's successor in 2014 and the subsequent drawdown of international forces.[17]

Determining Success

The case-study chapters focused on domestic actors look at the most crucial decisions and policies related to the establishment of democratic government and the rule of law and the state's approach to non-state justice. Because domestic actors almost always carry great influence, assessing external state-building efforts is immensely difficult (Fukuyama 2004). These challenges are compounded by the fact that consolidating the rule of law is always a prolonged process. Nevertheless, some efforts to promote the rule of law have positive effects, while others do not (Stromseth, Wippman, and Brooks 2006). An appropriate evaluation metric must recognize the complexity of promoting the rule of law and the extensive time required, as well as international actors' potentially important but invariably secondary role. Therefore, my research follows the logic for evaluation articulated by Paris in his seminal work on post-conflict state-building—namely, whether it has "enhanced the prospects for stable and lasting peace" (Paris 2004: 55). Because this research focuses on judicial state-building activities in states with high levels of legal pluralism, it assesses whether international aid has *enhanced the prospects for developing and consolidating the rule of law.*[18]

[16] An exit marks a meaningful decrease in the extent and nature of international involvement in state-building. Nevertheless, "external parties may, and often will, continue to be engaged in state building long after an operation has formally ended" (Caplan 2012: 5).

[17] While most external troops had departed by the end of 2014, the removal of international forces was repeatedly delayed until 2021.

[18] This criterion offers a significantly tougher test than the position articulated by Downs and Stedman. They judge activities solely on whether peace prevails when outside assistance departs (Downs and Stedman 2002: 50). Analyzing state-building, however, demands a serious examination of what occurs during and after a major period of international assistance (Paris 2004: 38–39, Tansey 2014). Most peace arrangements, even without outside assistance, can provide for a brief period of relative stability before a relapse into conflict. The other extreme is also problematic. Under the positive peace paradigm, as articulated by Galtung, success requires eliminating the "structural violence" facing ordinary people (Galtung 1969). Addressing structural violence demands tackling the harms done by "poverty and unjust social, political, and economic institutions, systems or structures" (Köhler and Alcock 1976: 343). By this standard, no state, let alone one that has recently experienced extensive violent conflict, meets this criterion.

As discussed above, given the immense challenges they face, it is most appropriate to assess post-conflict states for progress toward a "thin" rule of law. This evaluation standard involves looking at the positive and negative actions of programs and initiatives as well as their impact on the development of legal institutions and the broader rule-of-law culture. It considers the role of the institutions of democratic governance, which is a prerequisite for the rule of law. It also takes into account how domestic and international actors have engaged with non-state justice authorities by recognizing that non-state actors, while by no means the only factor, can dramatically influence judicial state-building efforts. This evaluation standard works well for societies featuring robust legally pluralist because it does not depend upon the state's institutional strength as a proxy for success. While important, international assistance is rarely, if ever, the primary determinant of the success or failure of judicial state-building, because it is invariably meditated through state officials and institutions.

Outline of the Book

This book consists of eight chapters. The first three outline the book's rationale and context. Chapter 1 examines my research agenda and methodological considerations. Chapter 2 analyzes key concepts related to promoting the rule of law in post-conflict legally pluralist states, including how the rule of law should be conceptualized, practical issues related to promoting the rule of law in post-conflict states, and terminology related to legal pluralism. Chapter 3 presents analytic archetypes for understanding the relationship between state and non-state justice systems. It argues that the relationship between the state and non-state justice sectors should be understood through a default archetype of competition after conflict. However, the situation is fluid. Domestic and international policy choices can help foster cooperative legal pluralism or trigger a descent into overtly combative legal pluralism. I propose five main strategies used by the state in relation to non-state justice systems: (1) bridging (2) harmonization, (3) incorporation, (4) subsidization, and (5) repression. There is no uniform template for success, but some strategies are better suited to certain environments. This chapter also analyzes the means frequently used to implement these strategies.

Chapters 4–7 are dedicated to the case studies of domestic and international judicial state-building efforts in Timor-Leste and Afghanistan.

Chapters 4 and 5 focus on Timor-Leste. Chapter 4 examines Timor-Leste's political history and institutions to show the foundation for both a modern democratic state and serious engagement with non-state justice. It also details post-independence efforts to build a modern state bound by the rule of law that works cooperatively with non-state judicial actors. Chapter 5 investigates conventional international efforts from 2002 to 2012 to promote the rule of law on the part of the United Nations, the United States, and Australia. On the whole, their contribution was generally positive, if modest. The chapter also looks at Portuguese efforts during the same era that sought to promote the rule of law while seeking to ensure the dominance of the Portuguese language. Portuguese assistance, while offering some benefits, raised serious concerns, because well over a decade after the end of Indonesia's occupation most Timorese could not understand the language.

Chapters 6 and 7 discuss Afghanistan. Chapter 6 surveys Afghanistan's history to show that a historical and cultural foundation exists for a legitimate, democratic state and legal order capable of supporting the rule of law that draws on long-standing pillars of legitimacy, namely Islam, tribal culture, and the provision of public goods. It then examines the post-Taliban political and legal institutions and highlights the new regime's decision to largely eschew these pillars and build a state dedicated to rent extraction rather than the rule of law. Chapter 7 critically examines international judicial state-building efforts in Afghanistan from 2002 to 2014. It demonstrates that not only were these efforts ineffective, there was never even a clear vision of how they could possibly ever be effective.

The final chapter reviews the major findings, highlights their scholarly and policy implications, and identifies areas for future inquiry.

2

The Rule of Law after Conflict

Introduction

The rule of law has an almost universal approval rating. It has been rightly recognized as cornerstone of liberal democratic states. Even regimes with little interest in liberal democracy have eagerly embraced the rule of law, including overtly authoritarian states, such as China (Whiting 2017); non-democratic states that still hold elections, such as Egypt (Moustafa 2014); and "competitive authoritarian" states where democratic structures still exist but are deeply compromised, such as Hungary (Kelemen 2020). The term "rule of law" can be used to describe everything from an authoritarian state justifying a brutal crackdown on peaceful protestors to citizens freely mobilizing to ensure government accountability in accordance with the law. "Rule-of-law" rhetoric can be invoked in widely different circumstances. And in non-democratic states, despite their claims to the contrary, the regime is usually not describing rule-of-law ideals at all.

At its core, the rule of law is not just about drafting laws and establishing legal procedures and getting people to comply with those laws. Nor is it simply about bolstering the capacity of the state judicial system. Any state can endorse these concepts. The rule of law is ultimately about beliefs and values, specifically the belief that everyone is bound by the same law and that the law protects and holds accountable both the governed and governing. Even a "thin" version of the rule of law must embrace equality under the law.

The rule of law and legal pluralism are complex, contested terms. This chapter explores different understandings of the rule of law and various manifestations of legal pluralism. Both terms tend to generate confusion regarding the relationship between normative and descriptive claims, albeit in different ways. A normative claim makes a value judgment about whether something is desirable or not, while a descriptive claim seeks to describe something as it is. Normative requirements are often omitted from discussions of the rule of law, which can be portrayed as a technocratic ideal where laws are enacted and enforced in a standard manner consistent with

Contending Orders. Geoffrey Swenson, Oxford University Press. © Oxford University Press 2022.
DOI: 10.1093/oso/9780197530429.003.0002

certain requirements. However, the ideal of the rule of law is inherently a normative one whereby state officials and society at large broadly commit to being bound by the law.

While the rule of law focuses on state-sponsored legal orders, in many places non-state justice enjoys more legitimacy and authority in people's everyday lives. Non-state justice, however, often involves elements that make outside observers uncomfortable. It can be tempting to use language to sanitize non-state justice processes. Non-state justice is often described in a way that contains normative elements that are not necessarily justified, such as being based purely on individual consent or always involving a neutral decision maker. This obscures the fact that non-state justice is always political, and it stands in the way of better understanding and more constructively engaging these systems by attempting to define away their potentially unsavory elements.

The first section of this chapter explores the concept of the rule of law, drawing a fundamental distinction between the rule *of* law and rule *by* law. Under rule *by* law, in authoritarian states or those with authoritarian tendencies, the law functions as a tool of regime authority and social control. Leaders and other privileged parties, however, are not required or committed to following its mandates. Under the rule *of* law, everyone, including state leaders, accepts being meaningfully bound by the law. With respect to the rule of law, this chapter explores thinner, more process-oriented understandings as well as "thicker," more substantive conceptualizations that include broader normative commitments. While both goals are worthwhile, I argue that a minimalist conception of the rule of law offers the most appropriate standard for assessing progress after conflict. The second section of the chapter addresses practical issues related to promoting the rule of law in post-conflict states. Although it is difficult to promote the rule of law in any society, let alone one devastated by war, the rule of law nevertheless forms a vital component of the democratic state-building process. Stepping away from the discussion of the rule of law and its promotion, the chapter's final section examines how legal pluralism should most accurately be understood and described. Various distinctions are commonly used in both academic and practitioner literature. The most common include formal and informal systems of justice, the state and customary systems of justice, the state and traditional systems of justice, and the state and non-state systems of justice. I argue that the distinction between state and non-state is the most accurate and logically coherent one for making generalizations and cross-case comparisons.

Understanding the Rule of Law

The idea of the rule of law is simultaneously technocratic, descriptive, and normative. In part reflecting this multidimensionality, the rule of law has retained a steadfast, though not uncontested, role as the structural foundation for stability, democracy, a market economy, and a plethora of other desirable institutions. The rule of law remains ubiquitous as the proposed "solution to the world's troubles" (Carothers 1998: 95). Promoting the rule of law can describe everything from improving economic performance to promoting good governance (May 2014). The rule of law's enduring, largely unchallenged status stands in marked contrast to that of liberal democracy, a concept with which the rule of law is frequently paired and in practical terms indelibly linked but which is currently facing serious ideological and practical challenges (Bermeo 2016).

Consensus about the rule of law's importance, however, comes at the expense of clarity. The literature abounds with seemingly endless definitions as well as distinctions across both history and geography (Jensen 2003; Kleinfeld 2006). Discourses about the rule of law, or ideas analogous to it, existed in classical Athens, ancient Rome, medieval Europe, and a wide variety of historical and contemporary societies (Tamanaha 2004). This ambiguity enables radically different rule-of-law ideals with starkly different outcomes in practice and can generate, intentionally or unintentionally, real confusion over what, exactly, the term means. Thus, it is important to consider its definition more systematically.

Amongst these various definitions, the common distinction between thick and thin understandings of the rule of law remains helpful (Peerenboom 2002b). At minimum, a thin concept of the rule of law requires that "law must be set forth in advance (be prospective), be made public, be general, be clear, be stable and certain, and be applied to everyone according to its terms" (Tamanaha 2007: 3). Even minimalist understandings of the rule of law are not purely technocratic, because "even the most formal, minimalist conception of the rule of law requires a normative commitment to the project of the rule of law itself" (Stromseth, Wippman, and Brooks 2006: 76). The rule of law, if it really is the rule of law and not mere rhetoric, must entail the idea that the state itself is bound by its law and views the law as a meaningful restraint on behavior. Indeed, state officials must be willing to be held accountable if they violate the law and must be willing to leave office if they are removed in accordance with duly constituted mechanisms for determining who is entrusted with the power of the state, ideally through open, competitive, and fair democratic elections—thus the important distinction between the rule *of* law and rule *by* law (Ginsburg and Moustafa 2008b). Going beyond minimalist understandings, thicker conceptualizations of the rule of law look to legal and broader political institutions, including compliance with international human-rights norms, media freedom, governmental checks and balances, and separation of powers (West 2003).

While a post-conflict state may strive for a more robust version of the rule of law, achieving a state of affairs approximating even a thin version is a major accomplishment. A state justice system must be able to make and implement binding rules across the territory it controls and on itself to uphold the rule of law. A state without the capacity to meet this lower threshold obviously cannot uphold more stringent or comprehensive definitions. Alternatively, as contemporary China highlights, a state can develop a robust, highly effective legal system but not make any meaningful progress toward the rule of law.

Regardless of the specific definition, the rule of law requires both the exercise of power and constraints on it. The state must be able to enforce its legitimate mandates. However, when states enhance the judicial system's capacity, their intentions are not necessarily benevolent. States may seek to codify regulations on public behavior, strengthen law enforcement, bolster the capacity and professionalism of the courts and legal personnel, and increase public awareness of the law to build either a more effective democratic state bound by the rule of law or a more effective authoritarian state consistent with rule by law. To be sure, codified, proscriptive law, a high-capacity justice sector, judicial independence, and public legal consciousness underpin governance systems committed to rule of law of ideals. However, these attributes are not the same thing as the rule of law. Authoritarian states may tolerate or even encourage a degree of judicial independence.

It is important to keep the rule of law as a clear standard and not conflate it with procedural reforms. In non-democratic states, judicial reforms can occur, which may actually improve the quality of justice for businesses and even ordinary citizens (Whiting 2017; Fu 2010). But initiatives that attempt to legitimize and professionalize an authoritarian regime are not the same thing as building the rule of law. The judiciary may emerge as an important ally for those resisting the regime, but even that cannot be assumed. Judiciaries can just as easily back authoritarian governments as oppose them

(Hilbink 2007). Insulating the judiciary and making it fully independent can be a huge boon for the quality of justice, but only if those judges act professionally and do not abuse that authority. A powerful, professional oversight body, for example, may help ensure that judges act with independence and professionalism in the interests of justice—or it can be a tool of political control.

Attempts by state officials to build the capacity of the judiciary and related institutions can support efforts to advance the rule of law but do not necessarily do so. The judiciary can both protect and oppress. The rule of law cannot exist when political elites are above the law. Establishing checks and balances and legal accountability for top leaders is always a challenge. As we will see in the next chapter, state attempts to increase judicial authority may be contested not only by those who previously operated with relative impunity, but also by the non-state justice providers that constitute the primary providers of order.

The second key step is even more challenging. The state, and more specifically those individuals and institutions who wield authority in its name, accept or are forced to accept restraints on their own power by being meaningfully bound by the law *even in areas they would prefer not to be*. Again, there can be substantial delegation or even some limits on government power in authoritarian or semi-authoritarian states. A competent judiciary with some real authority may serve the ends of a decidedly non-democratic regime. The judiciary may ratify controversial decisions by the executive or even mandate controversial reforms that regime may want to enact but also distance itself from (Moustafa 2007). Competent legal systems can increase the central government's ability to exert control over lower levels of government (Peerenboom 2002a). They can help resolve disputes between various state agencies or help enforce executive term limits to increase regime longevity and help reduce the threat of personalistic rule (Albertus and Menaldo 2012; Frantz and Stein 2017). Regimes may embrace procedural regularity and fairness (at least for "normal" disputes that do not threaten the regime's vital interests) to bolster legitimacy (Rajah 2012; Massoud 2013). States may develop a market-based economy that better protects property rights and promotes foreign investment through domestic or international law (Rajah 2012; Massoud 2014; Sidel 2008). In short, authoritarian or semi-authoritarian states may allow limits on their authority, but these constraints exist for strategic purposes and rarely if ever pose a threat to their hold on executive power. When the rule of law exists, state leaders are forced to follow the law even if it is inconvenient, counterproductive to their preferred policy outcomes, or detrimental to their political futures.

The rule of law is an ideal. In practice, even states generally viewed as bound by the rule of law engage in activities that contravene rule of law ideals. Take the United States, for example. The actions of former President Donald Trump frequently contravened rule-of-law ideals (Havercroft et al. 2018). Even former President Barack Obama, widely seen as having much more concern about upholding the rule of law, violated due process when he ordered the targeted killing overseas of a U.S. citizen (Blum and Heymann 2010: 151). When states seen as having something akin to the rule of law engage in behavior that contravenes that ideal, it serves as an important reminder that the rule of law can be threatened even where it is generally well established. The rule of law requires constant maintenance. Nonetheless, even though no state upholds rule-of-law ideals all the time, some states are closer to those ideals than others. The rule of law is therefore best understood as a spectrum ranging from closer to or further away from rule of law ideals rather than as a binary on/off situation whereby a state either possesses the rule of law or does not.

Legitimacy and the Rule of Law

The rule of law presupposes a legitimate state legal order, but not all legitimate orders feature the rule of law. Developing the rule of law is a prolonged process. Even under ideal circumstances, the rule of law takes roughly fifty years to develop (North, Wallis, and Weingast 2009: 27). The existence of a legitimate legal order, then, is necessary but not sufficient for establishing a state bound by the rule of law. Legitimacy exists "to the extent that its rules are considered rightful by both dominant and subordinate members of society" (Hechter 2009: 280). Legitimacy can be developed in various ways that do not require democratic government or the rule of law (Beetham 2013). Throughout history, tradition, military success, economic development, ideology, and provision of goods and services have all helped legitimize nondemocratic states. Compliance with legal stipulations reflects the popular belief that it is legitimate even if the state's democratic credentials are questionable or nonexistent (Tyler 2006), but that compliance is by no means inevitable even when the existence of the state itself is not subject to a serious challenge. This is particularly true in places marked by a high degree of legal pluralism (Migdal 1988). The provision of public services can, but does not necessarily, bolster a state's legitimacy (Mcloughlin 2015). Even states that enjoy substantial legitimacy may still face daunting capacity issues.

The Rule of Law and State-Building

The key question is not whether the rule of law has been established in the years following conflict. Almost inevitably, it has not. Rather, it is important to examine (1) whether a *legitimate* political and legal order has been created, and (2) whether judicial state-building has facilitated *progress* toward the rule of law. Assessments about whether progress toward building the rule of law has occurred should not be confused with determinations about whether the rule of law exists in the wake of conflict. For instance, independent Timor-Leste quickly established a high degree of legitimacy (Call 2008b: 1496) but still fell well short of even minimum rule-of-law ideals, as the quality of justice remained uneven and the main legal language, Portuguese, was not accessible to most people. Nevertheless, it had a legitimate political and legal order and made demonstrable progress toward the rule of law. After the end of the Taliban regime in 2001, Afghanistan likewise had the chance to foster a new, legitimate legal order under the auspices of a democratic state. Instead, the state constructed a vast network of rent extraction of which the legal system was a key component.

Law is vital to state-building. It underpins a modern nation-state's ability to project state-sanctioned regulations over its territory, gain international acceptance, and secure power both domestically and internationally (Ghani and Lockhart 2009; Paris and Sisk 2009). As the state's "authority is backed by a legal system and the capacity to use force to implement its policies" (Giddens 1993: 309), the rule of law rightly forms a major focus of both academic research and concrete state-building efforts. Praise has been coupled with serious investment. For instance, donors spent roughly US$2.6 billion on rule-of-law initiatives in 2008 alone (International Development Law Organization 2010).

Yet, the relationship between the rule of law and state-building can spark tensions. While the state can bolster its authority and legitimacy by operating in accordance with the law, such strict adherence to legal regulations may prevent leaders from taking necessary actions to benefit society at large (Fukuyama 2011: 246). This tension is endemic even in high-capacity states.

Societies clearly have an interest in seeing guilty individuals punished for criminal behavior. They also have an interest in preventing the state itself from breaking the law. This is the logic behind the long-standing but controversial exclusionary rule in the United States, which holds that in certain circumstances evidence that was collected illegally must be excluded from court proceedings. Individuals guilty of crimes may escape punishment because key evidence cannot be introduced at trial if the state failed to follow its own rules for evidence collection. The idea behind the rule is that it is sometimes more important to keep the state from engaging in illegal behavior than to ensure every possible conviction (Oaks 1970; Bradley 2012).

International efforts to strengthen a post-conflict state and facilitate the rule of law may create additional challenges. International support can weaken that state's ability to perform vital activities. Aid provision by international actors may be favored over local approaches, distort the local economy, establish suboptimal or even counterproductive incentive structures by injecting foreign capital into pre-existing and new patronage systems, and strengthen potential spoilers by paying for their (temporary) acquiescence (Mac Ginty 2008; De Waal 2015).

In practice, post-conflict states can have significantly different governance structures and different levels of international involvement, which has serious ramifications for how the rule of law is pursued. At one extreme lies autonomous recovery "through which countries achieve a lasting peace, a systematic reduction in violence, and post war political and economic development in the absence of international intervention" (Weinstein 2005: 9). Weinstein contrasts autonomous recovery with aided recovery, in which "international intervention plays a significant role in bringing war to an end, maintaining or guaranteeing a negotiated settlement, and assisting in the recovery process" (Weinstein 2005: 9). Each type of transitional regime presents unique challenges, opportunities, incentives, and limitations in promoting the rule of law. The cases examined in Chapters 4 through 7, Afghanistan and Timor-Leste, were both instances in which international actors played a significant, but not determinative, role under initial conditions of competitive legal pluralism.

Regardless of the type of transitional regime, promoting the rule of law in post-conflict states is no easy task. It requires enough laws to regulate behavior sufficiently, state officials willing to be bound by the law, courts that can actually adjudicate based on the law, competent legal professionals, law-enforcement organizations that actually enforce the law nationwide, and various auxiliary institutions (Fukuyama 2004: 59). Drawing on insights from organizational theory, Fukuyama notes that establishing the rule of law is "one of the most complex tasks state builders need to accomplish," as it is very difficult to ensure accountability since both law enforcement and dispute resolution involve very high numbers of transactions (2004: 59).

State political and legal institutions matter for establishing the rule of law, but even more is required. The rule of law has both top-down and bottom-up elements. From a top-down perspective, leaders must be committed to the rule of law. From a bottom-up perspective, citizens must demand that their leaders follow the law and believe in the ideals of a society bound by the rule of law. Respect for the law hinges largely on the widespread societal belief that the law is, at its core, basically fair and legitimate (Tyler 2006). Regardless of the state, "The rule of law is as much a culture as a set of institutions, as much a matter of the habits, commitments, and beliefs of ordinary people as legal codes" (Stromseth, Wippman, and Brooks 2006: 310). In post-conflict settings, popular faith in state institutions has inevitably been shaken, often shattered. Establishing the rule of law demands constructing and consolidating the popular legitimacy of new legal norms and institutions, just as much as it requires building courthouses or training judges. It is equally essential not to assume that these societies present "an institutional blank slate," since even in the most devastated countries institutions, customs, norms, and at the very least memories endure (Cliffe and Manning 2008: 165). Nearly all societies, no matter how devastated by conflict, possess the foundations of a possible rule-of-law culture. Often this means selectively drawing on preexisting beliefs and values, though it can be something as simple as promising a new order that offers a break from the lawlessness and violence of the past.

Promoting the Rule of Law in Practice

Promoting the rule of law after conflict means supporting some profoundly transformational goals. These include:

> Human security and basic law and order; A system to resolve property and commercial disputes and the provision of basic economic regulation; Human rights and transitional justice; Predictable and effective government bound by law; and Access to justice and equality before the law. (Samuels 2006: 7)

Achieving any of these goals—let alone of all of them—requires extensive reform of existing institutions and practices and often the wholesale construction of new institutions or a radical shift in the mentality and operations of existing ones.

Rule-of-law ideals are grand and abstract. Rule-of-law programs, however, tend to be modest and granular. This gap can be striking in international legal development assistance, which usually involves working with state and civil society institutions as well as other international actors. Despite repeated calls for reform, the basic activities favored by internationally supported rule-of-law initiatives have remained surprisingly static since the 1990s (Kleinfeld 2012).

So what does rule-of-law promotion actually entail? Generally, aid focuses on reforming institutions, rewriting laws, strengthening the legal profession, and improving access to the legal system (Carothers 1999: 168). After conflict, rule-of-law assistance often entails supporting the creation of new institutions or reestablishing old ones in environments "defined by the dual crisis of security and legitimacy" (Barnett and Zürcher 2009: 28).

Donor-funded initiatives tend to be relatively short-term and focused on concrete deliverables. International support, for example, may fund the drafting of legislation, underwrite the costs of operating justice institutions or the construction of buildings, pay for domestic and international staff, and sponsor training and educational materials.[1] Immediate, enumerable results appeal to donors and implementers alike, as they can demonstrate what specifically has been achieved with the allocated funds (Carothers 2006b). Beyond enumerating completed activities, even determining exactly what programs have accomplished toward their overarching goals remains a challenge. Despite ongoing criticisms of monitoring and evaluation approaches as ineffective, poorly designed, and uninformative, little progress is apparent in making high-quality monitoring and evaluation the norm rather than the exception (Cohen, Fandl et al. 2011; Clements 2020). Even when high-quality monitoring exists, evaluation faces fundamental challenges relating to the diffuse, complex nature of the rule of law. A program can deliver or even exceed all the intended outputs, but that does not guarantee that program activities will actually promote the rule of law

[1] Assistance to the legislature or executives, while usually relegated to a separate area, could arguably also be characterized as rule-of-law work. For the case studies, the intent of the implementers will serve as the primary criteria for determining whether such assistance counts as rule of law programming.

(Carothers 2003; Swenson 2017). Even in the case of program activities that clearly have had a positive effect, the most that can be realistically said about these efforts is that they made a modest contribution to advancing the rule of law. This is because establishing the rule of law is an extended, non-linear process beyond the timeframe of any particular initiative and invariably reflects domestic developments more than international aid. Contemporary evaluation is useful in its own right and as an important data source for scholars. At the same time, this book seeks to go beyond those evaluations. It offers a close, critical examination of the animating ideas, actions, and consequences of rule-of-law endeavors that examines domestic and international judicial state-building in Afghanistan and Timor-Leste within a broader political and historical context.

Defining Justice Outside the State

The state-sanctioned legal system is a pillar of the modern nation-state's domestic power and claim to territorial sovereignty in the international system. The state legal system's legitimacy and ability to resolve legal disputes are largely taken for granted in developed countries. It is difficult to fathom a developed country without a state-sanctioned legal system that has achieved a near monopoly over significant disputes. Yet, globally, this is the exception rather than the norm. Legal pluralism where multiple, distinct justice systems exist has a long historical pedigree (Benton 2002) and can be found everywhere, from highly localized communities to the international system (Berman 2012). As will be discussed further in the next chapter, the key distinction is to what extent alternative justice systems possess a meaningful degree of independence from the state. I thus define "non-state justice" as systematized quasi-legal processes that do not involve the state's implementation of its own laws. State laws themselves may be based on any combination of custom, tradition, religion, and/or a plethora of other sources. Still, a fundamental difference exists when those ideas are codified in formal law and state power is harnessed to encourage and sanction behavior.

Legal pluralism abounds even in states with highly developed state legal systems. Particularly at the lower levels, courts often rely on custom (Ellickson 1991). U.S. federal courts, for instance, regularly encourage parties to litigation to settle disputes through mediation prior to being allowed to proceed in the state system (Hensler 2004). Misdemeanors and even felonies may be settled through non-state means with the blessing of state officials. Arbitration even allows parties to agree to accept outcomes from a non-state actor as legally binding and explicitly agree not to follow state law (Dezalay and Garth 1998). Nevertheless, a high-capacity state retains the ultimate control, as other non-state dispute-resolution mechanisms must operate within state-sanctified confines, and their existence depends on staying in the state's good graces. While relatively systematized non-state dispute-resolution mechanisms exist outside the state system in even the most developed counties, the state largely retains the ultimate authority. In many developing countries, particularly post-conflict societies, however, this situation is reversed. Non-state legal systems handle the vast majority of disputes largely free of state interference (Kyed 2011).

The literature on legal pluralism distinguishes between state and other justice systems in various ways. The most commonly cited distinctions are between the formal and informal, the formal and customary, and the formal and traditional justice systems. Finally, the distinction between state and non-state legal systems also exists in the literature; I argue that this latter division both is logically sound and offers an internally coherent distinction that is useful for understanding legal systems across states. The next subsection begins with a discussion of general issues with the other above-mentioned pairings before going on to examine the advantages of distinguishing between non-state and state legal systems. The final two subsections delve more deeply into the problematic characterizations of non-state systems as informal and customary, respectively.

Non-State Legal Systems

Conceptualizing legal pluralism generates immense complexities and extensive scholarly disagreement (Woodman 1998; Tamanaha 2008). Yet, there is a shared recognition that state laws are not the only meaningful constraints on institutional and individual behavior, particularly in post-conflict states where state capacity is weak and state actors are often predatory against their own populations (Krasner 2004). In these circumstances, state legitimacy and resources are highly contested. Rules and regulations based on custom, tradition, religion, family lineage, and other sources of authority not sanctioned by law (such as criminal enterprises or powerful commercial interests) may hold greater sway than state law. Violations of these rules may lead to severe sanctions that include death. State authorities may even acquiesce in their use.

Classifying legal systems as either state or non-state offers substantial advantages. First, categorizing systems based on state or non-state status offers a meaningful and accurate distinction between different institutional mechanisms. Secondly, the state/non-state distinction is value-neutral regarding the body of law and system outcomes. Making a distinction between state and non-state avoids the linguistic baggage associated with terms such as "informal," "traditional," and "customary," which involve empirical and often normative claims. The state/non-state distinction facilitates inquiries regarding system operations and outputs without presuppositions regarding those findings. It also avoids being drawn into complex definitional questions such as "How long does a system have to be in place before it qualifies as traditional?" or "How much structure can exist in an informal system?"

The state/non-state distinction also benefits from parsimony and renders additional qualifications, such as the specification of a "formal state system," unnecessary, at least at a general level. For example, Wojkowska's influential 2006 UNDP report states, "The *formal justice system* for the purposes of this paper involves civil and criminal justice and includes formal state-based justice institutions and procedures, such as police, prosecution, courts (religious and secular) and custodial measures" (Wojkowska 2006: 9, italics in original). A particular state's justice system may be rather formal or informal in practice. Definitions along these lines say little more than that the formal justice system is the state system; they say nothing about how the system actually operates in its own right or relative to other dispute-resolution forums. Thus, "formal" does little to illuminate these contested concepts as an independent description or as an additional qualification to the term "state legal system."

While the state/non-state distinction is not particularly prevalent in the policy literature, it is commonplace in the legal-pluralism literature (for example, see Griffiths 1986; Connolly 2005; U.N. Commission on Legal Empowerment of the Poor 2008; Baker 2010a; Faundez 2011). For example, a United Kingdom Department for International Development (DFID) working paper notes the comprehensive nature of the term "non-state justice," which "includes a range of traditional, customary, religious and informal mechanisms that deal with disputes and/or security matters" while avoiding potentially unwarranted assumptions about those systems (DFID 2004: 1). Scholars and practitioners recognize the term's utility, and while most do not use the state/non-state distinction exclusively, few would argue that the distinction is inappropriate.

Not surprisingly, reality often complicates the clear division between state and non-state actors. Individuals and institutions frequently act in ways that blur the distinction (Baker 2010a). For example, police in Timor-Leste may return cases to the non-state authorities even when victims, especially women claiming they have experienced domestic violence, seek to access state courts, and even though police are legally required to facilitate entrance to the state justice system (Everett 2009). Prominent people may simultaneously hold positions of authority within the state bureaucracy and the non-state legal order. Furthermore, the state may incorporate certain elements of non-state law into state regulations, or even the entire system (Connolly 2005). For example, Vatican City's and the Islamic Republic of Iran's legal systems are overtly predicated on religious norms. Yet even there not all religious rules are state law. State officials retain discretion over which religion-based regulations are backed by state coercive power and which are left to individual or communal discretion. Thus the distinction between state and non-state legal systems holds up even in extreme cases, including when a state is expressly religious in outlook or draws expressly on custom for many of its laws. While these instances no doubt generate ambiguity at the border, they do not fundamentally challenge the state/non-state distinction. Moreover, as the next section highlights, using terminology other than "state" and "non-state" legal systems as terms for cross-case comparisons presents real problems.

Overarching Issues

Before moving on to the more specific issues raised by the framing of legal systems as informal, customary, or traditional, it is worthwhile to examine some common cross-cutting issues relating to how legal pluralism is described in its own terms and in relation to the state.

Omission

Despite the centrality of legal systems that enjoy meaningful autonomy from the state, authors often fail to define key terms such as "custom," "formal," "informal," "indigenous," and "traditional." These terms are by no means self-explanatory or uncontested, so it is essential to clarify them: arguments about legal pluralism that do not include clear definitions lack a solid foundation. While characterizing legal systems as "customary," "informal," and "traditional" is problematic for making comparisons across countries, particular non-state systems in specific locations may in fact best be described using those terms. However, it is vital to know exactly what authors mean when they use each term to determine whether or not it is appropriate.

Conflation and Confusion

Scholars and practitioners frequently use the terms "non-state," "indigenous," "customary," "traditional," and "informal" interchangeably. However, all of these have a wide variety of possible connotations and different procedural and substantive implications. Because each of those terms has potentially problematic elements, causally interchanging the terms magnifies their uncertainty and problematic aspects.

Normative Smuggling

Characterizing a legal system as "customary," "indigenous," "traditional," or "informal" almost invariably introduces normative claims regarding that system. At best, these claims are questionable and require empirical evidence. More often they are misleading and, on occasion, simply wrong. The inclusion of normative stipulations in functional definitions likely stems at least in part from a recognition of non-state justice systems' more problematic aspects. In addition, the terminology may also reflect a desire to simultaneously endorse those processes while avoiding association with their more problematic elements. A major U.N. system report is illustrative. It defines informal justice systems as involving "a neutral third party not part of the judiciary" while also noting that "custom-based systems appear to have the advantages of sustainability and legitimacy" (United Nations Development Program, U.N. Women et al. 2012: 8). Yet, there is no reason that an informal system would require a neutral actor (even if one were to grant the highly questionable assumption that decision makers can be neutral) let alone intrinsically sustainable and legitimate. Indeed, because non-state systems reflect local power structures, it is unrealistic to expect a neutral decision maker (Swenson 2018b).

These terms may obscure or downplay the coercive underpinnings of non-state justice systems. For example, an International Development Law

Organization (IDLO) report argues that customary law is completely consensual because "customary law is only law to the extent people follow it" (Harper 2011a: 17). An effective customary legal system cannot require consent from all parties. After all, if the system relies on total consent to make its rules binding, then the system is not a legal system. Legal systems seek to constrain behavior deemed undesirable, and at a minimum they require authority to mediate competing interests and some capacity to sanction rule violators (Raz 2009). These claims may be accurate in certain cases, but they must be proven rather than assumed. In contrast, the state/non-state distinction is value-neutral and does not presuppose any normative process or output.

Informal Legal Systems

The term "informal justice" is also popular. Distinctions are generally drawn between the "formal" state justice systems and "informal" dispute-resolution mechanisms, which exist outside the state (Benton 1994; Wojkowska 2006; Chopra 2008; UNDP et al. 2012). For example, the aforementioned U.N. study defines informal justice systems as "encompassing the resolution of disputes and the regulation of conduct by adjudication or the assistance of a neutral third party that is not a part of the judiciary as established by law and/or whose substantive, procedural or structural foundation is not primarily based on statutory law" (UNDP et al. 2012: 8).

Despite its popularity, the term "informal" remains ambiguous. Definitions tend to focus on the informal justice system's otherness with respect to the state. Even when authors are not doing so explicitly, they often define the informal justice system as consisting of those subsystems that do not fall under the umbrella of the state. The core concept seems to be that the informal system is less routinized and more flexible than the state justice system. This notion, however, is often misleading, for informal systems can be highly formalized. For example, many ethnic Pashtuns in Afghanistan draw on a non-state system called Pashtunwali that is known for its complexity, formality, and comprehensiveness. In contrast, a state legal system can draw upon state law while being highly ad hoc procedurally or in the application of substantive law (Bierschenk 2008). Government officials may disregard state law in practice or may not even know the relevant law. Depending on the context, state legal systems can function informally, and non-state legal actors can behave with a high degree of formality.

Customary and Traditional Legal Systems

"Customary law" is used to identify boundaries based on observed behavior within a community rather than statutory or codified law (Glenn 2011: 42). An IDLO report reflects a common assumption about the customary legal systems from which customary law originates:

> customs, norms and practices that are repeated by members of a particular group for such an extent of time that they consider them to be mandatory. Customary systems tend to draw their authority from cultural, customary or religious beliefs and ideas, rather than the political or legal authority of the state. As such, provided that it has not been incorporated into state law, customary law is only law to the extent that the people who follow it, voluntarily or otherwise, consider it to have the status of law. (Harper 2011a: 17)

By definition, whatever else it entails, customary law requires a degree of routinized behavior, imbued with meaning and with a sense of history, even if that understanding is not technically accurate (Anderson 2006). However, this is not the case for many legal systems deemed customary. Paradoxically, so-called customary law tends to be "dynamic, adaptable and flexible, and any written version of it tends to become outdated quickly" (Harper 2011a: 17). This understanding of customary law is rather paradoxical, for it envisions a system that simultaneously changes rapidly while embodying the voluntary, well-established, broad-based consensus of an entire community.

Customary law relies on normative ideas about what a community does and what communities should do. Definitions of customary law often assume its legitimacy, when in fact the legitimacy of any legal system, whether state or non-state, cannot be taken for granted—it must be determined on the basis of evidence. Defining customary law systems as running on a purely voluntary basis obfuscates reality and ignores the often extensive coercive power wielded by key actors and institutions. Moreover, customary justice systems involve power dynamics that may reflect a broad-based social agreement locally but still conflict with international norms. Even observers who see virtues in customary justice recognize it is often biased against women, for example, and that the quality of justice received can vary dramatically by status (Chopra and Isser 2012; United Nations Development Program, U.N. Women et al. 2012).[2]

Finally, the term "customary law" fails to cover the broad range of non-state legal systems present in post-conflict societies. Legal pluralism assumes an almost bewildering variety of forms. By constructing a largely binary relationship between state and customary legal systems, the portrait of legal pluralism in a given setting can quickly become distorted.

The state legal system is also frequently contrasted with notions of traditional justice (Hohe and Nixon 2003; Clark 2007; Schmeidl 2011). The term "traditional justice" is also often used as a synonym for "informal" or "customary" (Stromseth, Wippman, and Brooks 2006; Isser 2011b). Like the term "customary," "traditional" carries normative implications, including the idea that such practices have meaningful roots within communities and enjoy a certain legitimacy. A constant is implied with state structures that may be new, imported from an external source, predatory, or all of the above. However, "tradition," like "custom," inevitably involves plurality rather than uniformity. In societies subject to conflict or other major social upheaval, structures deemed "traditional" might take on new, innovative forms that bear only a passing resemblance to their institutional predecessors. While the legitimacy of traditional processes is often assumed, they remain subject to internal and external pressures that make imputing legitimacy as an inherent quality problematic.[3] For example, shortly after the fall of the Taliban in 2001, "political armed groups, commanders, and militias have strategically targeted traditional and customary justice systems in some parts of rural Afghanistan in an attempt to exert control over local populations" (Nojumi, Mazurana et al. 2004: 35).

In short, the term "non-state justice" generates less confusion and is better for cross-case comparison. Other, related terms can be used constructively in specific circumstances, but it must be shown empirically that any such term truly fits.

[2] Although women frequently endure systematic discrimination in state courts, this discrimination is generally an aberration that goes against the state law's requirements rather a consequence of following the accepted rules (Campbell and Swenson 2016).

[3] The same caveat applies to the state justice system, but the idea of the state system lacks similar normative implications.

Conclusion

Advancing the rule of law is major goal of post-conflict state-building. The rule of law is a meaningful, albeit contested, concept. It is deeply linked with democratic governance. The rule of law requires democracy, but not all democratic states enjoy the rule of law or are even making progress toward it. Bolstering state law, legal bureaucracy, and law enforcement are necessary but not sufficient for establishing the rule of law. A state can have a powerful, effective, and even legitimate legal system and still not be committed to the rule of law unless state officials themselves are bound by the law. The rule of law also provides a benchmark against which actual efforts can be judged. Legal pluralism is a complex, multifaceted phenomenon understood and described in a wide variety of ways. However, for general cross-case comparisons, the distinction between state and non-state legal systems is the most useful. With these clarifications in mind, we can turn to a discussion of different manifestations of legal pluralisms and how to conceptualize judicial state-building efforts in legally pluralist environments.

3

Understanding and Engaging Pluralist Legal Orders

Introduction

The previous chapters highlighted the vital importance of post-conflict judicial state-building. Yet, the state justice sector does not exist in a vacuum. All states feature legal pluralism, but the prevalence, autonomy, and authority of non-state justice systems vary dramatically. Only a limited number of high-capacity states have non-state justice actors firmly under their control.[1] Even in these states, legal pluralism thrives through alternative dispute-resolution mechanisms, arbitration agreements, and international treaty obligations. In the developing world, most disputes are handled outside the state justice system (Kyed 2011). The most authoritative rules may not be state regulations. Custom, tradition, religion, family lineage, and other influences not sanctioned by law (such as those of criminal enterprises or powerful commercial interests) may have an influence equal to or even greater than that of state law. The role of legal pluralism is particularly vital in conflict and post-conflict settings, as they tend to have weak state institutions and contested governing authority (Isser 2011b). In states with less capacity and/or legitimacy, seeking support from non-state judicial actors can serve as a conflict-avoidance tactic or even a broader governance strategy that attempts to secure buy-in from powerful, potentially antagonistic groups.[2]

[1] Mainstream international relations theory, for example, tends to simply assume a unitary state with a monopoly on the use of legitimate violence domestically; however, the reality of political and legal authority is far more complicated. As Lake highlights, "the state is central to the study of international relations and will remain so for the foreseeable future" because it is "fundamental" to neorealist, neoliberal institutionalism, and constructivist approaches, and usually plays a key role in "critical, post-modern, and feminist theories" (Lake 2008: 41).

[2] These strategies are by no means limited to post-conflict states. In Malaysia, for instance, the longtime ruling party has used state-sponsored Islamic courts as an integral part of its efforts "to validate its Islamic credentials—relegitimize the party and the state—and thus co-opt, or at least undercut, both the Islamic resurgents and the opposition party" (Peletz 2002: 10).

Contending Orders. Geoffrey Swenson, Oxford University Press. © Oxford University Press 2022.
DOI: 10.1093/oso/9780197530429.003.0003

Although the legitimacy and endurance of many non-state justice forums can be seen as an indictment of the need for state justice to be underpinned by the rule of law, non-state justice mechanisms often have significant negative externalities. Non-state legal orders frequently reflect cultural or religious norms unconcerned with basic human rights. Women and other vulnerable groups are particularly at risk when non-state legal systems embrace overtly patriarchal ideals. These systems can be heavily biased toward powerful individuals and families, and legal processes often lack core procedural and substantive due-process protections. As Waldorf highlights, non-state " 'judicial' elites are neither independent nor impartial, and their discretionary rulings serve community harmony, not individualized justice" (Waldorf 2006: 10). Furthermore, the relationship between state and non-state justice is often unclear, and cases may be resolved in different ways, encouraging the parties, particularly those with more economic or political clout, to forum shop. The state system's predominance in itself does not guarantee a basically just legal system, as it could be a means for more effective despotism. Nevertheless, a competent justice system is a functional prerequisite for a state capable of fulfilling the rule of law's requirements of being prospective, generalized, clear, fixed, and applied equally (Tamanaha 2007: 3).

While central governments prefer a monopoly on legal authority, state judicial power is frequently contested long after conflict officially ends. Domestically, the state's ability to function justly and effectively is a matter of life or death for millions of people worldwide and dramatically impacts the quality of life for millions more. Internationally, "poorly governed societies can generate conflicts that spill across international borders" as well as facilitate criminal networks and transnational violent extremism (Krasner 2004: 86). Establishing a viable state justice sector is vital to the overall success or failure of state-building efforts (Paris 2004: 205–206). Because non-state justice is the dominant form of legal order, it is important to understand its implications. State-building provides a powerful analytical lens to examine and understand the implications of legal pluralism, as situations tend to be fluid, with a wide range of realistic outcomes.

Robust legal pluralism challenges the state's claim to a monopoly on legitimate resolution of legal disputes as well as the ideal of uniform application of the law. It enables participants to select dispute resolution forums based on accessibility, efficiency, legitimacy, jurisdiction, and cost, as well as the state and non-state systems' respective abilities to make binding decisions and sanction individuals who choose other systems. This process leads to a sustained struggle between state and non-state justice actors for legitimacy, resources, and authority. Apart from their adjudicatory role, non-state justice authorities can even become challengers to the state itself and help spark renewed violence. On the other hand, seeking support from non-state judicial actors can be vital for stability and securing elite and popular support for the state.

Legal pluralism matters everywhere, but it matters far more in some states. It is not enough to merely recognize that legal pluralism exists; scholars and policy makers must understand how legal pluralism actually functions. This chapter explores legal pluralism theoretically and empirically in three main sections. The first highlights important existing conceptual work on the relationship between state and non-state justice while also demonstrating the need for new archetypes and clarity regarding the application of various strategies. The second section posits four distinct archetypes of legal pluralism for relations between state and non-state justice: (1) combative, (2) competitive, (3) cooperative, and (4) complementary. Vibrant legal pluralism and the relationship between state and non-state systems directly influence the results of international and domestic efforts to build state legal capacity and promote the rule of law after conflict. Although variations in legal pluralism do not fully explain why cases turn out differently, the type of legal pluralism present and how domestic and international actors engage with it play a crucial role in aiding or hindering the building of the rule of law. The chapter concludes by illustrating commonly used strategies for interacting with non-state justice actors across different legally pluralistic environments. These strategies are always present even if sometimes unstated. There are five main strategies for conceptualizing engagement between state and non-state systems: (1) bridging, (2) harmonization, (3) incorporation, (4) subsidization, and (5) repression. Different environments favor different strategies. The best prospect of success in promoting the rule of law requires an understanding of the legal pluralism archetype and its programmatic implications, along with credible and appropriate strategies for transforming the current environment into a more constructive situation.

In subsequent chapters, case studies from Timor-Leste and Afghanistan will highlight how both positive and negative changes in the overarching relationship between state and non-state justice can occur.

Understanding State and Non-State Justice Systems

Conceptualizing the relationship between state and non-state justice systems has received relatively little scholarly attention. More generally, however, Helmke and Levitsky (2004) have produced a widely cited typology for understanding interactions between formal and informal institutions. The typology evaluates interactions using two criteria. The first criterion examines whether formal and informal institutional outcomes are convergent or divergent based on "whether following informal rules produces a substantively similar or different result from that expected from a strict and exclusive adherence to formal rules" (Helmke and Levitsky 2004: 728). The second criterion assesses whether formal institutions are effective or ineffective by looking at "the effectiveness of the relevant formal institutions, that is, the extent to which rules and procedures that exist on paper are enforced and complied with in practice" (Helmke and Levitsky 2004: 728). The typology has four main descriptions of formal-informal institutional interaction: complementary (convergent, effective formal institutions), accommodating (divergent, effective formal institutions), competing (divergent, effective formal institutions), and substitutive (divergent, ineffective formal institutions) (Helmke and Levitsky 2004: 726).

Helmke and Levitsky's typology explores the different ways political actors create and utilize informal institutions, defined as "*socially shared rules, usually unwritten, that are created, communicated, and enforced outside of officially sanctioned channels*" (Helmke and Levitsky 2004: 727, italics in original). They explain: "We restrict our analysis to the modern period, when codification of law is nearly universal" (Helmke and Levitsky 2004: 726). Their approach assumes a state with a relatively high degree of legitimacy and effectiveness and avoids an overtly antagonistic relationship with informal institutions. In short, it assumes an overlap of actors in formal roles with those in informal roles. This certainly can be the case, but it is not necessarily so. While highly useful in its own terms, Helmke and Levitsky's framework differs in both analytical approach and goals from my own.

This project is distinct in four major ways. First, my approach goes beyond examining issues of "convergence and effectiveness" between informal and formal institutions by exploring legal orders that rest on different normative foundations. This dynamic is quite common owing to the highly contested nature of legal orders in many developing countries, particularly post-conflict ones. It is not useful to analyze Taliban justice in Afghanistan, for example, in terms of its convergence and effectiveness in meeting state goals. After all, as discussed extensively in Chapters 6 and 7, the Taliban legal system sought to destroy and supplant the Islamic Republic of Afghanistan's legal system.

Second, law is not a major focus for Helmke and Levitsky, and their typology does not fully capture the distinct manifestations of legal pluralism. They acknowledge legal pluralism (Helmke and Levitsky 2004: 725) and note briefly that "scholars of legal pluralism have argued that the imposition of European legal systems created 'multiple systems of legal obligation'" (Helmke and Levitsky 2004: 729). Because these systems "embodied very different principles and procedures, adherence to custom law at times required a violation of state law (and vice versa)" (Helmke and Levitsky 2004: 729). This constitutes their entire discussion of legal pluralism. My theoretical approach is novel because it explores the various types of legal pluralism that exist across all settings.

Third, Helmke and Levitsky are concerned with informal institutions rather than state versus non-state actors and institutions. This book consciously avoids using the terms "formal" and "informal" as categories to describe justice systems. This may seem a minor distinction, but it is a vital one, at least with regard to legal systems. As I have argued elsewhere:

> Non-state justice is often referred to as informal, traditional, or customary law. However, these terms might not capture the empirical reality. Informal systems can, in practice, be highly formalized. Ethnic Pashtuns in Afghanistan draw on a non-state system based on longstanding cultural beliefs, Pashtunwali, known for its complexity, formality, and comprehensiveness. On the other hand, the state legal system can be highly ad hoc, and state officials may disregard or may not even know the relevant law. Rather than drawing an unhelpful distinction between formal and informal/traditional/customary, classifying justice as either state or non-state offers substantial advantages. The state/non-state distinction is value-neutral regarding content and outcomes. It avoids the linguistic baggage associated with terms such as "informal," "traditional," or "customary," which inherently involve empirical and often normative claims. (Campbell and Swenson 2016: 115–116).

Fourth, Helmke and Levitsky do not discuss informal institutions that fundamentally challenge the state or actively seek to undermine it. In contrast, my theory recognizes that legal orders are often contested, especially in developing countries. In situations with multiple legal systems, even in complementary legal pluralism, there are almost always divergent normative foundations. Even in a country such as the United States, arbitration clauses in contracts are explicitly designed to circumvent, not strengthen, relevant federal law. More broadly, Helmke and Levitsky assume a relatively stable, high-capacity state. They seek to understand "why, given the existence of a set of formal rules and rule-making mechanisms, do actors choose to create informal rules?" (Helmke and Levitsky 2004: 730). In other words, their approach assumes a level of state-building success that does not necessarily exist in practice. Indeed, it is what this book seeks to explore: how states create a legitimate state-based legal order in places where that order is contested, most notably post-conflict states. My approach for analyzing state and non-state justice actors and institutions has the advantage of covering the full spectrum of states, ranging from those where state and non-state legal orders are violently opposed to those where they peacefully coexist.

Finally, Helmke and Levitsky acknowledge legal pluralism but equate law with other types of service provision. In contrast, I contend that the legal system is different because it speaks directly to one of the core characteristics that makes a state sovereign in the international system: the monopoly on the legitimate use of violence. The legal system is by far the most prevalent mechanism for states to legitimately apply violence toward their own citizens. Thus, law is qualitatively different from other public services.[3]

[3] There is some overlap in the use of the terms "complementary" and "competing," but the two have different meanings. Helmke and Levitsky use the term "complementary" primarily to refer to the belief and cultural structures that underpin state institutions, whereby "complementary informal institutions may also serve as a foundation for formal institutions, creating or strengthening incentives to comply with formal rules that might otherwise exist merely on paper" (Helmke and Levitsky 2004: 728). They emphasize these "informal institutions do not merely exist alongside effective formal ones, but rather play a key role in making effective the formal rules of the game" (Helmke and Levitsky 2004: 728). I use "complementary" to describe situations where both state and non-state justice institutions exist, but non-state justice mechanisms operate under the purview of state authority. However, this does not require that non-state venues support the state venue in any way. For example, the arbitration agreements noted above are commonplace under complementary legal pluralism as I have described it, but their function is not to "make effective the formal rules of the game" (Helmke and Levitsky 2004: 728). Instead, arbitration aims to avoid or even undermine many of the rules imposed by the state. Likewise, Helmke and Levitsky use the term "competing" in the sense of institutions that are divergent and ineffective because "formal rules and procedures are not systematically enforced, which enables actors to ignore or violate them" (Helmke and Levitsky 2004: 728). Again, it describes different phenomena from those in my own theory. Competitive legal pluralism refers to the existence of multiple legal systems that rest on divergent normative foundations and actively compete for users, rather than a failure to enforce formal rules.

Other scholars from a range of disciplines have generated important insights. Since Malinowski's pioneering work, anthropologists have highlighted the existence and operations of functional non-state legal orders. In Trobriand, Malinowski argued, law "consists . . . of a body of binding obligations, regarded as a right by one party and acknowledged as a duty by the other, kept in force by a specific mechanism of reciprocity and publicity inherent in the structure of their society," rather than state-based regulations (Malinowski 1985 [1926]: 58). Moreover, anthropologists have offered compelling insights into the construction of, and resistance to, non-state authority (Moore 1986, 1993; Weiner 2006). And research has illustrated diverse and complex ways legal pluralism has related to state and transnational legal orders within state boundaries and beyond them (Wilson 2000; Zips 2005; von Benda-Beckmann, von Benda-Beckmann, and Griffiths 2009).

Addressing the challenge of legal pluralism is not simply foundational to the establishment of the modern state; it predates the state. The recognition of competition between the state and non-state justice systems dates back to the seminal work of the legal historians Pollock and Maitland (1923). They note the chronic tensions among multiple legal systems in England and the state's desire to expand its judicial power, at the expense of customary systems. Despite a very high degree of legal pluralism, they demonstrate how "the custom of the king's court" became "the custom of England," as over time "the central power . . . quietly subsumed all things into itself" (Pollock and Maitland 1923: 184–186). Drawing on their work, Fukuyama (2011: 245–275) contends that competition between legal systems drove the formation of many modern states. Legal pluralism also shaped interactions between different societies, and "dual legal systems were widespread in colonized parts of Africa, Asia, Latin America, and the Pacific" (Merry 1991: 890). Legal pluralism became a defining feature of colonial administrations that sought to harness local dispute resolution mechanisms to help legitimize and institutionalize their rule (Benton 2002; Renders 2012; Balint et al. 2020). Likewise, multiethnic domains, such as the Ottoman Empire, embraced legal pluralism (Barkey 2013). Nor is the prevalence of legal pluralism largely a historical phenomenon. Currently, legal pluralism exists everywhere in forms ranging from community dispute resolution to the international system, where there has been a proliferation of treaties as well as transnational regimes with veto capacity or even legislative ability (Berman 2012). State and non-state legal systems can work together relatively smoothly or find themselves clashing frequently (Tamanaha 2008).

Other scholars have emphasized the vital role of justice systems in the success or failure of state-building endeavors (Paris 2004: 205–206; Rubin 2008). Beyond formal institutions, Cao demonstrates the vital role that culture plays in advancing and consolidating the rule of law (Cao 2016). In most post-conflict settings, non-state mechanisms resolve an overwhelming majority of disputes (Wojkowska 2006; Kyed 2011; Baker 2010b). There are some existing typologies for understanding the interactions between the state and non-state justice sectors. Connolly proposes the approaches of abolition of non-state systems or alternatively incorporation, partial incorporation or no incorporation of non-state mechanisms (Connolly 2005: 247–249). Forsyth argues that the relationship between state and non-state justice mechanisms should be seen as a seven-stage "spectrum of increasing state acceptance of the validity of adjudicative power by the non-state justice system" (Forsyth 2009: 202). Taking a different approach, Loveman seeks to explain how the state increased its administrative capabilities and jurisdiction by looking at state strategies of innovation, imitation, co-option, and usurpation in relation to non-state actors (Loveman 2005). While stopping short of offering a full typology, numerous scholars have noted some potential permutations when state and non-state justice systems interact (Tamanaha 2008; Baker 2010a; Isser 2011a). Although these approaches can be quite helpful for understanding state efforts to engage non-state justice mechanisms, they offer limited insights into post-conflict judicial state-building efforts in states with extensive legal pluralism.

Moreover, direct application of existing models to state-building efforts are insufficient because they fail to take into account the particular challenges of post-conflict settings (Call and Wyeth 2008; Paris and Sisk 2009). Successfully asserting and maintaining authority domestically and internationally invariably challenges even relatively high-capacity states (Tilly 1992; Fukuyama 2014). As Krasner and Risse note, "While no state governs hierarchically all the time, consolidated states possess the ability to authoritatively make, implement, and enforce decisions for a collectivity" (Krasner and Risse 2014: 549). In post-conflict settings, state power is almost always actively contested. In post-conflict settings where there are significant international post-conflict state-building efforts, the state's power is almost certain to be meaningfully contested (Baker and Scheye 2007).[4] It is wrong to

[4] In contrast, after the withdrawal of their armed forces from Iraq, the United States, the United Kingdom, and other coalition partners were not generally considered to be post-conflict states even though they had participated in a conflict that had ended. The post-conflict states that are the target

assume that the state can simply impose its will on non-state actors (Migdal 1988; Faundez 2011). Existing typologies overwhelmingly focus on state strategies rather than the strategies of international state builders. State and international strategies may overlap because international state builders often attempt to partner with state actors. Sometimes international state builders even directly control the machinery of the state (Caplan 2005) or select the local elites who will exercise power (Manning 2006). Ultimately, however, the archetypes and strategies proposed below can help to contextualize all state-building efforts regardless of whether they are undertaken by local elites under the auspices of the state, international state builders, or some combination thereof.

State and Non-State Justice Relations Archetypes

This chapter proposes a new typology more attuned to legal pluralism's structure and implications that provides a richer theoretical understanding of how legal pluralism functions in all states, as well as its relationship to the rule of law. As discussed in the last chapter, the concept of the rule of law is used in a "thin" rather than "thick" sense (Peerenboom 2002b). At minimum, a thin concept of the rule of law requires that "law must be set forth in advance (be prospective), be made public, be general, be clear, be stable and certain, and be applied to everyone" (Tamanaha 2007: 3) as well as a commitment to rule-of-law ideals by state officials and society more generally. The institutional, economic, cultural, and political requirements for thicker understandings of the rule of law, while they are important long-term goals, are unrealistic for most post-conflict states in the short or even medium term.

Establishing even a thin version of the rule of law after conflict is slow and difficult. State action alone is not enough. The rule of law requires formal and informal mechanisms for popular accountability. Building the rule of law is inherently a fluid, contested process that includes both top-down and bottom-up elements. These dynamics are particularly acute in post-conflict settings because the public trust in state institutions has almost inevitably been undermined, often severely. In practical terms, the rule of law requires

of international intervention discussed in this book have almost invariably seen conflict as destructive, at least in terms of state capacity, rather than constructive, even if the conflict has dislodged a deeply unpopular previous regime.

consolidating the popular legitimacy of new legal norms and institutions at least as much as it demands competent professionals, physical infrastructure, or revised legislation.

This section posits new theoretical archetypes for understanding the fluid relationship between state and non-state justice in a wide range of settings: (1) combative, (2) competitive, (3) cooperative, and (4) complementary. Furthermore, it proposes and examines five main strategies for understanding international and domestic judicial state-builders' engagement techniques: (1) bridging, (2) harmonization, (3) incorporation, (4) subsidization, and (5) repression. These strategies illuminate the main domestic and international approaches to engaging the non-state legal sector in legally pluralist societies.

The strategies discussed here are top-down in the sense that they examine how domestic and international actors can attempt to proactively influence the overarching type of legal pluralism in a given setting. In practice, however, the overarching legal-pluralism archetype invariably reflects top-down and bottom-up factors. Whether non-state actors choose to engage the state and on what terms is decidedly bottom-up, as is to what extent and under what conditions people use the state and non-state justice systems. As with other areas of state policy, popular mobilization and political advocacy is a potential tool to change the state's approach to non-state justice (Jordan and Van Tuijl 2000). For instance, bottom-up activities could include popular efforts to end repression of non-state justice actors and institutions or alternatively through political advocacy against recognizing legal dispute systems that are seen as violating human rights. Therefore, both the overarching archetype of legal pluralism present at a given moment and movement between different types of legal pluralism reflect a mixture of top-down and bottom-up elements.

In practical terms, the strategies help inform which activities can help foster change from one archetype to another, as well as how advances in judicial state-building could be achieved within a given archetype. Rather than viewing the process as linear, advancing the rule of law after conflict should be conceptualized as highly fluid, with points of progress and regression. Even in favorable circumstances, establishing the rule of law and the inclusive institutions that underpin it is usually a prolonged, highly contingent process with both top-down and bottom-up elements (North, Wallis, and Weingast 2009; Acemoglu and Robinson 2012). Nevertheless, understanding how legal pluralism functions at a macro level can significantly

Table 3.1 Legal Pluralism Archetypes

Archetype	Key Features	Examples
Combative	The state and non-state justice sectors do not recognize each other's right to exist and actively seek to destroy each other.	Afghan state from 2004 to 2021 with Taliban justice system
Competitive	Deep tensions exist between the state and non-state justice sectors, and there are frequent clashes between systems. However, the state's formal juridical authority is not challenged. While the non-state justice sector retains autonomy, the state and non-state systems respect each other's right to exist in some form.	Afghanistan from 2001 to 2003; Afghan state with tribal authorities from 2004 to 2021; East Timor from 1999 to 2002; Jordan under the Hashemite monarchy
Cooperative	The non-state justice sector retains a significant degree of authority and autonomy. However, state and non-state legal authorities are generally willing to work together toward shared goals.	Timor-Leste after independence, particularly since 2006; Zimbabwe after 1980
Complementary	Both state and non-state justice exist, but non-state justice mechanisms operate under the umbrella of state authority.	United States, United Kingdom, Germany, Japan

bolster domestic and international initiatives designed to influence the non-state justice sector and minimize the risk of costly missteps.[5]

While invariably a simplification, archetypes help conceptualize the core features of the relationship between state and non-state judicial actors (see Table 3.1). They can inform judicial state-building efforts as well as other domestic and international policy initiatives that influence or can be influenced by non-state actors, such as economic development initiatives or broader governance endeavors. Although the case studies here focus on post-conflict settings, the archetypes can help actors understand and improve their engagement in a wide variety of legally pluralist states. Whatever the context, by understanding the dominant archetype, it is

[5] These missteps can take a wide variety of forms and can involve both domestic and international actors. Isser highlights that state policies attempting to regulate the conduct and jurisdiction of non-state systems in Liberia and Guatemala "have undermined the effectiveness of TAs [traditional authorities] without providing an alternative outlet" (Isser 2011a: 333).

possible to engage constructively with the non-state justice sector in a given area.[6] Constructive engagement with the non-state justice sector requires thinking critically about how to deal with current realities while simultaneously seeking to transform them.

Combative Legal Pluralism

In combative legal pluralism, state and non-state systems are overtly hostile to one another. The state and non-state justice sectors seek explicitly to undermine, discredit, supplant, and, ideally, destroy the other. Combative legal pluralism can involve non-state actors rejecting the state system's ideological foundation in a largely nonviolent manner. For instance, in the fight against the apartheid state in South Africa, there were active efforts to establish "structures of justice and policing that contested the legitimacy of their equivalent in the apartheid institution" (Nina 2000: 24). While compatible with nonviolence, combative legal pluralism often flourishes in countries facing an active insurgency or separatist movement. Combative legal pluralism flourishes in countries facing an active insurgency or separatist movement. Indeed, non-state justice often forms a significant foundation of those movements (Kalyvas 2006: 218–219; Mampilly 2011; Arjona et al. 2015). The importance of justice systems is not merely academic. Che Guevara's treatise *Guerrilla Warfare* stresses the need to establish a rival justice system:

> In view of the importance of relations with the peasants, it is necessary to create organizations that make regulations for them, organizations that exist not only within the liberated area, but also have connexions in the adjacent areas. . . . The peasants will sow the seed with oral and written propaganda, with accounts of life in the other zone, of the laws that have already been issued for the protection of the small peasant, of the spirit of sacrifice of the rebel army; in a word, they are creating the necessary atmosphere for helping the rebel troops. (Guevara 1969: 93–94)

[6] Relationships may vary by location as well as by religious, tribal, or ethnic group. The state may enjoy a cooperative relationship with non-state actors in one part of the country while facing a competitive or even combative relationship in other areas. For example, the Islamic Republic of Afghanistan had a competitive relationship with many tribal and religious non-state authorities but a combative relationship with the Taliban justice system.

Unsurprisingly, combative legal pluralism abounds where post-conflict state-building and efforts to promote the rule of law have failed or are clearly trending in a negative direction. In many instances, combative legal pluralism coincides with an active insurgency characterized by parallel antagonistic state-building enterprises, and the scope for even limited collaboration is minimal to non-existent (Wickham-Crowley 1992; Kalyvas 2006: 218–220). For example, as Chapter 6 explains, in Afghanistan the Islamic Republic and the Taliban justice systems sought to destroy each other (Johnson 2011: 282; Coburn 2013). The state refused to recognize Taliban adjudication and attempted to eliminate Taliban judicial personnel. People in Taliban-controlled areas were barred from accessing state courts and even faced collective punishment for engaging them (Johnson and Mason 2007; Giustozzi and Baczko 2014).

Competitive Legal Pluralism

Under competitive legal pluralism, the state's overarching authority is not challenged, but non-state actors retain substantial autonomy. This dynamic characterizes many developing countries and is extremely common after conflict. Indeed, it is the default condition in post-conflict settings. While a post-conflict political and legal settlement has been reached, it has not yet been consolidated or institutionalized. The prospect of a return to conflict is disturbingly common (Collier, Hoeffler, and Söderbom 2008). The state invariably finds itself trying to assert a new order in places previously beyond its control or in places where that control was contested.[7] It is also common in many developing counties where societal actors retain autonomy and still exercise the right to maintain order in a way not dictated by state officials (Migdal 1988). Competitive legal pluralism features significant, often deep, tensions between state and non-state legal systems, especially where legal norms and procedure diverge significantly. Yet, in these situations, clashes seldom endanger the state's formal judicial authority, because the non-state justice sector does not seek to fully supplant state authority. While the non-state justice sector retains a sizable degree of autonomy, the state and

[7] This friction between legal systems is by no means a purely modern phenomenon, as almost every successful judicial state-building endeavor has historically required the state to suppress, outperform, or collaborate with non-state rivals (Fukuyama 2011).

non-state systems respect each other's right to exist in some form and are willing to engage with each other, at least tactically (Baker and Scheye 2007).[8]

Competitive non-state legal systems are most frequently rooted in religious beliefs or shared culture, custom, or heritage. These traditions can be real or invented (Anderson 2006). Such systems of order often exist outside the state and do not necessarily share the state's values.[9] Competitive legal pluralism also exists where criminal actors have established illegal markets that rarely seek to take over the state but actively work to retain autonomy by undermining and circumventing state law to establish a separate, parallel order with enforcement capacity (Volkov 2000; Skarbek 2011). There may be collusion between state officials and criminal actors to evade the law or profit illegally; however, the activities themselves remain fundamentally opposed to state law (Uslaner 2008; Rothstein 2011).

In post-conflict settings the level of competition tends to track the success of the state-building process. For example, post-conflict state-building efforts in Afghanistan and Timor-Leste both started against a backdrop of competitive legal pluralism. In Afghanistan, where state-building trended in a decidedly negative direction that helped spur renewed violent conflict, competitive legal pluralism has slipped into combative legal pluralism. Alternatively, when the state gains legitimacy, authority, and capacity, the environment and the incentives change. This is what occurred in Timor-Leste. Non-state actors increasingly favor collaboration because both the advantages of partnership with the state and the disadvantages of opposing it increase. Nonetheless, as long as non-state actors retain a high degree of autonomy, the potential for setbacks is present.

Competitive legal pluralism can be prolonged and endemic, particularly when non-state dispute resolution is seen as legitimate and authoritative. In relatively stable states, it can even be an explicit strategy.[10] These difficulties are compounded in post-conflict settings where the state's legal authority and institutions have been severely compromised. While states often seek to

[8] This dynamic is common but not inevitable. Winning a conflict can kickstart the process of establishing a legitimate state legal order that supplants or undermines non-state actors, but this process takes time.

[9] The fact that many of those traditions or cultural beliefs may be invented or imagined does little to undermine their power when people share an ideological commitment to the ideas they embody (Anderson 2006).

[10] For example, during the period of stability from 1923 to 1978, Afghan rulers used just such a strategy. The state demanded the allegiance of tribal leaders and the local population. Local politics and the resolution of local disputes, however, were left largely to local populations (Barfield 2010: 220).

supplant non-state competitors over time, in other circumstances engagement with non-state authorities forms the basis of the governance system. Domestic judicial state builders often rely on powerful, established non-state actors and structures to support their rule, including but not limited to tribal or clan groups and religious authorities (Migdal 2001).

In Afghanistan, for instance, competition between legal systems is deeply embedded. Since the state's inception, the Afghan government "sought to impose centralized code-based judicial institutions" (Barfield 2008: 349). The state recognized that non-state justice remained the dominant form of dispute resolution. In return, the state required religious and tribal authorities to acquiesce to the regime (Gregorian 1969; Rubin 2002). These tensions were usually managed but occasionally exploded into open, violent conflict, such as when religious and tribal authorities cooperated to depose King Amanullah in 1928 (Poullada 1973). As highlighted in Chapter 6, even the Taliban's rapid imposition of a state legal order in the 1990s reflected a tacit agreement with prominent forms of tribe-centric justice. Many years later, support from, and influence with, tribal and religious authorities again formed an important part of the Taliban's prolonged insurgency and eventual return to power in August 2021.

Cooperative Legal Pluralism

In a cooperative legal-pluralist environment, non-state justice authorities still retain significant autonomy and authority. Non-state judicial actors, however, have by and large accepted the state's normative legitimacy and are generally willing to work together toward shared goals. Major clashes are far less frequent and tend to focus on social issues where values clash, such as women's rights, rather than existential issues of state judicial power. Cooperative legal pluralism tends to thrive in places where progress is being made toward consolidating legitimate state authority. This shift may coincide with meaningful advances toward the consolidation of democratic governance bound by the rule of law.

At the same time, it is important to stress that cooperative legal pluralism is cooperative only in terms of the relationship between the predominant forms of state and non-state justice. Establishing a legitimate state that enjoys a cooperative relationship with non-state authorities does not necessarily require democracy or the rule of law. This dynamic is reflected by post-conflict transformations in places such as Zimbabwe (Kriger 2006). Cooperative legal pluralism does not necessarily mean that, in terms of substance, the law is just. A legal order characterized by this archetype invariably still produces winners and losers. Under cooperative legal pluralism, the law, whether state or non-state, can still be used to violate human rights, oppress citizens, or perpetrate systematic discrimination against certain groups. Alternatively, a state justice system does not need to be particularly effective to enjoy significant legitimacy provided the overarching state is legitimate (Kriger 2003, 2006). Yet cooperation has its limits. In many postcolonial African states, non-state justice actors largely backed the nascent state, but their support often waned "when the state-building project [became] too exclusivist or predatory" (Dorman 2006: 1087).

By 2012 in Timor-Leste, state and non-state justice actors pursued a common agenda focused on stability and development even as tensions over women's rights lingered (Wassel 2014). This constitutes a remarkable accomplishment given the state's limited capacity and the continued prevalence of non-state justice (Marx 2013). It was made possible by a combination of historical factors, institutions, and strategic choices. Timor-Leste's state, with support from the international community, transformed a legal context long dominated by competitive legal pluralism under Portuguese colonial rule, which had become combative under Indonesia's occupation, into a legal landscape distinguished by cooperative legal pluralism. Non-state authorities now view the state and the state-building endeavor as legitimate. They are largely open to constructive engagement because state leaders were also recognized as key actors in the struggle for independence.

Complementary Legal Pluralism

Legal pluralism does not disappear in a state with a high-capacity, effective legal system, but it becomes complementary. Non-state justice is subordinated and structured by the state because the state enjoys both the legitimacy to have its rule accepted and the capacity to actually enforce its mandates (Ellickson 1991). The United States, Western Europe, and many other countries with high-capacity legal systems choose to allow private arbitration, mediation, and other forms of alternative dispute resolution (ADR) (Cappelletti 1993; Nolan-Haley 1996; Gaitis 2004; Redfern, Hunter et al. 2004). ADR takes various forms, but generally "a third party is involved who offers an opinion or communicates information about the dispute to the disputants" (Shavell 1995: 1). For civil disputes, state courts frequently mandate claimants' attempts to settle their disputes outside of court before being allowed to move their case forward (Stipanowich 2004). However, substantively and procedurally, state and non-state law can still clash. Arbitration agreements facilitate the evasion of state law and legal process, but the extent of circumvention depends on the policy preferences of state officials. In all instances, these processes are integrated into, and fall under the ultimate regulatory purview of, the state and exist at its pleasure, and they largely depend on state courts for enforcement. ADR processes are allowed and often encouraged because the state deems them useful for addressing real and perceived "inefficiencies and injustices of traditional court systems" (Edwards 1986: 668).

This form of legal pluralism is complementary from a governance perspective because the state has effectively outsourced alternative forums, such as court-referred mediation, or at least tactically licensed dispute venues, such as binding arbitration. Complementary legal pluralism features a similarly cooperative ethos, but the non-state justice authorities operate under the umbrella of state authority and without substantial autonomy to reject state decisions. Only complementary legal pluralism can truly uphold that oft-stated requirement for the rule of law: that the law is applied equally to all people (Carothers 1998; Tamanaha 2007). Complementary legal pluralism is a worthwhile long-term goal, but it is important to have reasonable expectations about what is feasible in the short to medium term after conflict. As with cooperative legal pluralism, complementary legal pluralism refers only to the nature of the relationship between state and non-state justice. As Organski and Kugler highlight, "A high-capacity state need not be free, democratic, stable, orderly, representative, [or] participatory" (Organski and Kugler 1981: 72). These types of legal orders *can* uphold the rule of law, but that does not mean that states with complementary legal pluralism necessarily *do* uphold the rule of law.

Strategies for Addressing the Non-State Justice Sector

While the predominant legal pluralism archetype is important, it is not immovable. As the case-study chapters highlight, it is possible to transform settings marked with high degrees of competitive legal pluralism into a cooperative environment, as occurred in Timor-Leste after independence and has so continued there. Alternatively, the situation can deteriorate from competitive legal pluralism to combative legal pluralism, as occurred in Afghanistan during the Islamic Republic. Thus, effective engagement with the non-state justice sector requires thinking critically about how to deal with current realities while simultaneously seeking to transform the environment into one more favorable to judicial state-building.

Domestic and international state builders have a number of strategies with the potential to promote significant change. Of course, the results of those efforts will reflect domestic or international actors' ability to persuade, or in some cases coerce, non-state authorities and society at large. As with other areas of policy, there is no guarantee that all organs of the state will pursue a consistent, coordinated approach to non-state justice. While justice policy is often dealt with at the national level, in states with a degree of decentralization, local or regional government may pursue distinct policy approaches (Benjamin 2008). Decentralized units' ability to pursue distinct approaches will vary according to their levels of authority. Nevertheless, state actors have the same strategies available whether or not they are actually able to pursue them.

As shown in Table 3.2, this section identifies five main strategies for how best to understand interactions between state and non-state systems: (1) bridging, (2) harmonization, (3) incorporation, (4) subsidization, and (5) repression. These strategies are by no means mutually exclusive or hermetically sealed, but they are nevertheless conceptually and functionally distinct. These strategies may have a stated end goal or vision, but this is not necessarily the case, as the assumptions may be implicit. Domestic and international state builders may employ these strategies with little regard for the complex relationship between the state and non-state justice systems, which can lead to decidedly suboptimal outcomes. The same donor may even fund multiple programs with divergent strategic approaches, as occurred in Afghanistan. This need not be the case. Success cannot be guaranteed, but certain strategies are better suited to certain environments. Savvy strategic planning and pragmatic adaptation, ideally combined with a bit of good fortune, has a reasonable chance of improving the relationship between the state and non-state justice sectors regardless of the dynamics present.

Table 3.2 Non-State Justice Sector Strategies

Strategy	Key Features	Examples
Bridging	Bridging strategies seek to ensure that cases are allocated between the state and non-state justice systems as appropriate based on state law, participants' preferences, and venue appropriateness.	Seeking to ensure serious crimes cannot be resolved outside state courts regardless of the disputants' preferences by using paralegals to direct cases to state courts or offering training on how to access state courts
Harmonization	Harmonization strategies seek to ensure that the non-state justice systems' outputs are consistent with the state system's core values.	Laws to outlaw discriminatory practices in non-state adjudication and training to end discriminatory practices
Incorporation	Incorporation strategies eliminate the distinction between state and non-state justice. Non-state justice, at least in a formal sense, becomes state justice.	Outcomes of the non-state justice systems are endorsed but also regulated by the state system. In practice, incorporation could mean the creation of explicitly religious or customary courts with state support or the labeling of non-state justice venues as state courts of first instance.
Subsidization	Subsidization strategies support the state judicial system to increase its capacity, performance, and appeal relative to non-state alternatives.	Facilitating legislative reform, establishing physical infrastructure used by the justice sector, supporting symbolic representation, capacity-building, and promoting public engagement
Repression	Repression strategies seek to fundamentally undermine and ideally eliminate the state's non-state rivals.	Outlawing non-state justice forums or seeking to arrest or kill non-state justice actors

Bridging

Bridging strategies seek to ensure that cases are allocated between the state and non-state justice systems as appropriate, based on state law, participants' preferences, and venue suitability. Almost invariably, bridging asserts jurisdictional claims regarding both state and non-state venues. Certain legal matters, most notably homicide and other serious crimes involving physical harm, must be resolved in state courts, while minor nonviolent disputes are almost always left to non-state venues. Public-information campaigns are frequently undertaken to increase public knowledge about the state legal system. Similarly, training can inform local leaders and citizens about the state legal system and how to access it. Free or subsidized legal aid can give people an economically viable choice between legal system systems. A bridging strategy can be paired with a formal incorporation approach that seeks to provide a state legislative framework for non-state justice, but it can also be a stand-alone initiative that seeks to increase individuals' choices without trying to resolve larger questions regarding the relationship between state and non-state justice venues.

Bridging can work well where unmet demand exists for access to state courts or when increased awareness stimulates demand for state justice. A bridging strategy's impact also hinges on non-state authorities' willingness to facilitate or accept referrals to the state system, which by extension decreases their autonomy. Thus, bridging is frequently a useful approach in competitive and cooperative legal pluralistic environments, though it offers little utility where the state has a combative relationship with non-state authorities. For example, bridging worked well in Timor-Leste, where non-state authorities willingly limited their jurisdiction and facilitated access the state system for certain disputes (RDTL Ministry of State Administration and the Asia Foundation 2013).

Harmonization

Harmonization attempts to ensure decisions of the non-state justice system comport with the state's core values. At the same time, non-state justice is accepted and legitimatized to some extent. To support harmonization, states and international donors often fund activities to encourage practitioners of non-state justice to act in a manner consistent with state law in general. However, there is often at least tacit recognition that non-state authorities retain a significant autonomy and legitimacy. There is a willingness to tolerate some normative differences in adjudication standards. As opposed to trying to get non-state venues to act like state courts of first instance, the focus is on changing how certain legal matters are handled—for example, the treatment of women in non-state forums (Chopra and Isser 2012). State judicial actors also frequently discriminate against women, but usually this is done in violation of state law rather than as accepted practice (Campbell and

Swenson 2016). In general, the greater the state's ability to offer a compelling and legitimate forum for dispute resolution worth emulating, the more likely harmonization strategies are to succeed. Successful harmonization occurs most frequently in competitive—and especially cooperative—legal pluralism environments. Nevertheless, as long as non-state actors retain a significant degree of autonomy, meaningful divergence with state policy remains possible.

This dynamic holds true even in settings such as Timor-Leste, where cooperative legal pluralism now exists. Both the central state and the international community invested heavily in eradicating domestic and other forms of gender-based violence. Local responses varied, but the treatment of domestic violence cases remained an acute area of tension between state and non-state legal orders. Historically, "Under the Indonesian law as well as in the traditional system . . . domestic violence was a purely private matter" (Hohe 2003: 345). Domestic violence is now unequivocally a public crime under the Penal Code (RDTL 2009b: Arts. 146, 154) and the Law against Domestic Violence (RDTL 2010b). Despite suco chiefs' general willingness to support the state's goals and uphold state law, domestic violence was still overwhelmingly treated as a private matter to be addressed by the non-state justice system (Wigglesworth 2013). The persistence of non-state dispute resolution for domestic violence is not particularly surprising, because the classification of domestic violence as a public crime contravened previous state and non-state practice (Corcoran-Nantes 2009; Cummins and Guterres 2012). How suco leaders address domestic violence remains largely discretionary when there is a meaningful difference between the domestic law and the suco leader's authority. This highlights how non-state authorities still retained an effective veto over state regulations in their jurisdiction even while generally supportive of the state and its law.

Incorporation

Under incorporation, the distinction between state and non-state justice is eliminated, at least from the state's perspective. Non-state justice, in a technical sense, becomes state justice. The non-state justice systems' decisions are endorsed but also regulated by state officials. Incorporation can mean the creation of explicitly religious or customary courts with state support and oversight, the labeling of non-state justice forums as courts of first instance, or simply by offering an avenue for appeal from non-state venues to state courts. Decisions of the non-state system could be subject to appeal or ratification by the state system. For example, a local council's decision regarding a property dispute could be formalized through a district court or administrative entity. At its extreme, the entire non-state justice system could be brought under the state justice system's purview (Peters and Ubink 2015).

While non-state systems are allowed to continue and perhaps even grow, in practice incorporation is a bold move to assert practical and ideological authority over non-state actors by limiting their independence. Judicial state builders seek to both harness the authority of the state system and control it. As Kyed and Buur have argued regarding post-conflict states in sub-Saharan Africa, governments often recognize non-state authorities with the aim of "restoring state governance and reextending state governance" (Kyed and Buur 2007: 5). After all, an incorporation strategy becomes necessary only if the state cannot achieve dominance locally. Once incorporated, states may further seek to regulate "customary" non-state law by codifying it, an inherently subjective and selective process, or even by creating new non-state law. Scholars have long recognized that "the very attempt to write down custom in a way meant to be juridically or intellectually definitive loses custom" (Fitzpatrick 1984: 21). Incorporation strategies may also be less overt through professional regulation or the induction of elections to non-state posts previously allocated through other means, such as ancestral lineage.

There is no guarantee that non-state actors will be willing or able to be incorporated into the state justice system. The state may envision itself as the principal, with the non-state judicial actors as its agent. This approach is highly problematic, however, if the non-state actors have notably different norms and values from the state and accountability mechanisms are weak. Thus, the prospects and effectiveness of incorporation strategies track the state's ability to persuade or compel large numbers of non-state judicial actors to engage with it. This enabling dynamic is most likely in cooperative settings, possible but tentative in competitive scenarios, and very unlikely in combative environments.

Subsidization

Subsidization of the state system is the most common strategy and has a vast number of potential targets. Unlike harmonization, bridging, and incorporation, subsidization does not require any meaningful acquiescence by non-state judicial actors. Consequently, under subsidization approaches, the non-state system is largely left alone or at least is not the primary target. The state system receives assistance to increase its capacity, performance, and appeal relative to the non-state system. This process sounds straightforward, but the tasks that must be accomplished are often extremely challenging. For instance, it is easy to fund trainings that seek to promote judicial independence, but it is hard to establish an independent judiciary.

Subsidization can take an immense variety of forms and can be implemented regardless of whether the environment is marked by cooperative, combative, or competitive legal pluralism. Subsidization is by far the most common strategy in post-conflict settings and in general. It is also most likely to influence the overarching relationship between the state and non-state justice systems because it is the only form of aid directed at improving the performance, legitimacy, and effectiveness of state justice. Certain key subsidization techniques tend to recur across settings, notably legislative reform, capacity building and establishing physical infrastructure used by the justice sector, supporting symbolic representation, and promoting public engagement.[11]

Legislative Reform

Predicated on the straightforward idea that drafting new laws and reforming existing ones can improve the justice system, legal reform is a major focus of subsidization efforts (Channell 2006; Stromseth, Wippman, and Brooks 2006; Jensen 2008). Drafting a new constitution or reforming the existing one is usually a top priority after conflict (Elkins, Ginsburg et al. 2007), as is crafting institutions to implement the new legal order or reinforce the existing one (Horowitz 2006). These areas are given particular attention because they are supposed to structure the state's legal and political order and offer a clean break from a violent past. Yet problems cannot simply be legislated away. While states often enact legislation rapidly, they have generally been slower, or even unable, to implement it. Implementation really matters because the capacity to enforce law is essential to idea of law, let alone the

[11] This list is by no means exclusive. For a more detailed description of activities, see Kleinfeld (2012).

rule of law (Raz 2009). Typically, such legislation upholds international standards and reflects principles and norms widely accepted in international advisors' home countries. Yet, it often clashes with local realities and beliefs. Even when the state justice system produces results consistent with the codified law, the courts and other legal actors often find implementation lacking or arbitrary. However, passing legislation or ratifying international treaties does not necessarily advance the rule of law unless the state is willing or compelled to be bound by the law (Hathaway 2002; Ginsburg and Moustafa 2008a).

Physical Infrastructure

Prolonged conflict invariably takes a toll on the state's physical infrastructure (Cliffe and Manning 2008). Judicial state builders frequently seek to build or repair courthouses, government ministries, and other buildings related to the legal system. Physical infrastructure is often prioritized based on the assumption that constructing viable spaces for the state justice sector to perform its duties leads to increased use and legitimacy. While workspace can be helpful, it does little to improve the quality of justice unless state officials have the knowledge, capacity, and desire to improve the justice system.

Symbolic Representation

Because the rule of law rests in significant part on the belief of both justice sector actors and the general population in the idea that the rule of law exists, symbolism, and the legitimacy it helps project, is vital (Přibáň 2007). Recognizing the role of symbolism, state builders may seek to enhance the representational power of the state justice system. Symbolism is deeply imbedded in any legal system, from architecture to court procedure and decorum (Tobe, Simon et al. 2013). The foundational legal instruments are also often venerated (Elkins, Ginsburg et al. 2009). The perception of legitimacy is absolutely essential because the judiciary lacks the authority to implement its decisions alone (Tyler 2006) and must be viewed as legitimate by elites and ordinary citizens alike (Fukuyama 2011: 247). When they are effective, legal "symbols do not merely generate feelings of unity" but also function as "a source of political mobilization" (Smith 1997: 935). The effectiveness of symbolism largely hinges on the underlying entity's legitimacy as well as how compelling the symbols are for the audience. A symbolic representation that enhances the state's legitimacy to one person can be decidedly exclusionary to another observer (Scott 1998; Horowitz 2000: 216–219).

Capacity-Building

Capacity-building is immensely popular, particularly with international actors, as it seeks to instill the skills and values necessary to establish the rule of law (Carothers 1999: 163–177). Technical assistance and educational investment for current or future legal professionals likewise forms a common approach to state-building in post-conflict societies. Yet the actual impact of training for building capacity, let alone establishing the rule of law, is notoriously difficult to demonstrate (Carothers 2003). Assistance often embodies norms that may not easily fit within existing culture and produces results that are hard to quantify. Nonetheless, the rationale behind capacity-building remains prominent, even though results are mixed because improving the knowledge and capacity of those operationalizing the state justice system is necessary even if insufficient on its own.

Public Engagement

Public engagement strategies can seek to improve people's understanding of the state justice sector and how to use it. Other aid is explicitly advocacy oriented, addressing issues such as political and judicial or legislative reform designed to improve how the judicial system functions (Diamond 2008: 126–133). These mobilization efforts may promote vital reforms. In other instances they can focus on areas that are of particular interest to the international community but possess limited domestic constituencies. For example, efforts may focus on international human-rights treaty obligations rather than relevant domestic legislation (Fowler 2013: 3).

Programs frequently veer into the realm of public relations by attempting to improve the popular opinion about the state justice system. These efforts are particularly challenging and potentially even harmful when they seek to increase traffic to state courts and improve the courts' reputation absent a corresponding increase in the quality of the justice system. For example, both major USAID programs in Afghanistan that focused on the state system—the Afghanistan Rule of Law Project and the Rule of Law Stabilization Program-Formal—tried to enhance the justice sector's public image and increase use of state courts (Checchi and Company Consulting 2009b; Leeth, Hoverter et al. 2012: 6). Improving the public's perception, however, did not fix the entrenched corruption within the Afghan judicial system (Singh 2015).

Pushing more people to use a predatory, highly corrupt state justice system does not advance the rule of law. It simply threatens to further undermine faith in the regime.

Repression

Repression strategies seek to fundamentally undermine, and ideally eliminate, the state's non-state rivals. Repression strategies are not concerned with persuading or incentivizing non-state justice actors to work with state authorities. Nor does repression hinge on the state system's persuading or influencing its non-state rivals or encouraging citizens to use state courts. Repression can simply be a matter of outlawing non-state justice forums, particularly in relatively peaceful places such as Botswana, and using the state's power to enforce its mandate (Forsyth 2009). Where the state can effectively outlaw non-state justice, however, the state is already predominant. Almost invariably repression involves significant violence, rather than merely the threat of sanction. As the state seeks to eliminate non-state justice actors, repression often results in reciprocal violence by non-state actors. Alternatively, violence by non-state actors can trigger state repression efforts.

In Afghanistan, for example, international and Afghan forces invested heavily in attempts to destroy the Taliban justice system through targeted killings and other projections of force (Giustozzi, Franco, and Baczko 2013; Giustozzi and Baczko 2014). The desire to repress is not solely the purview of the state. The Taliban sought "to discredit and undermine" state claims to legal authority by impeding the work of state courts and engaging in assassinations of state legal personnel (Johnson 2011: 282).

Repression efforts are rarely paired with bridging, harmonization, and incorporation strategies because those approaches depend on constructive engagement with non-state authorities. However, the state frequently subsidizes the state justice system in an attempt to increase its authority and effectiveness relative to non-state justice. It can also include efforts to protect state judicial officials and infrastructure from insurgent attacks, as the United States tried to do in Afghanistan (SIGAR 2009, 2015). While invariably unpleasant in practice, repression can be an important tool when the state faces an existential threat from non-state justice actors, particularly when those actors are linked to an armed insurgency. Nevertheless, repression alone, even when backed by force, is unlikely to be sufficient for the state to consolidate a monopoly on legal authority, since force must be paired with another form of legitimacy to be sustainable over time (Beetham 2013).

Conclusion

Understanding legal pluralism is important for any legal or policy intervention, including but by no means limited to state-building. Without an understanding of legal pluralism's dynamics in a given context, interventions are likely to be ineffective. Even initiatives that enjoy short-term success are unlikely to be sustainable, because they reflect good fortune rather than an informed approach. Sound strategy requires understanding how state and non-state actors interact systemically. This chapter helps build that knowledge by presenting the four main types of relationships between state and non-state actors through a typological framework that illuminates the dynamics of legal pluralism across contexts. If we understand what legal pluralism archetype currently exists, an appropriate strategy or package of strategies for engaging with non-state actors can be selected. By identifying the main strategies available to policy makers, this chapter illustrates how each approach works within each archetype and which strategies might be appropriate to deploy depending on the context. Such strategies must be rooted in a deep understanding of a country's culture, politics, and history along with a keen understanding of the potential foundations for a legitimate legal order. Realistic expectations and time horizons are also essential, because developing the rule of law takes a long time.

Because competitive legal pluralism is the default setting after conflict, it is worth examining states that employ it in more detail. Thus, the subsequent case-study chapters focus on this most common condition in post-conflict states, rather than trying to highlight examples from each archetype. At the onset of judicial state-building efforts after war, states are very rarely post-conflict while still being characterized as having combative legal pluralism. After all, combative legal pluralism is usually a sign of continued or renewed conflict. In contrast, after conflict tensions are still fresh, and powerful non-state actors have found themselves on both sides of the conflict (Staniland 2012). It is unlikely that the nascent post-conflict legal order will initially enjoy a cooperative, let alone complementary, relationship with all major non-state judicial actors. The rare exceptions tend to be places where a legitimate, high-capacity state that featured complementary legal pluralism existed prior to the conflict and one side of the conflict has experienced a clear victory that has been generally accepted by the former combatants. For example, post–World War II Germany and Japan would meet these criteria (Dobbins et al. 2003).

Advancing the rule of law depends primarily on domestic actors and institutions. The international community, however, can offer incentives (or disincentives) for actions by both state and non-state justice actors. Policy makers benefit from being better at conceptualizing and responding to the challenges posed by a legally pluralist society. Constructive engagement with non-state legal authorities is particularly vital when they are strong enough to successfully resist state or internationally led initiatives, as is generally the case in post-conflict and conflict-prone settings. Thus, it is crucial to understand the overarching archetype of legal pluralism to inform the specific strategies that are being used. The strategies highlighted above have been used by domestic and international state builders in reference to the non-state justice sector. These techniques have been widely replicated across contexts, but rarely critically examined. The following chapters seek to remedy this deficit by investigating domestic and then international judicial state-building endeavors in Timor-Leste and Afghanistan.

PART TWO
CONTENDING ORDERS AFTER CONFLICT

Post-conflict situations are fluid. Progress toward the rule of law, or lack thereof, needs to be viewed holistically. It involves the legal system but also the overarching governance structures related to the law. While certainly not the only factor, non-state justice plays a crucial, though generally underappreciated, role in influencing the success or failure of both domestic and international judicial state-building efforts. Because post-conflict settings can feature a wide range of relationships between state and non-state actors, they are particularly instructive for showing how archetypes of legal pluralism can be shifted over time. Case studies from Timor-Leste and Afghanistan highlight the fact that selecting an appropriate policy is vital to achieving sustainable positive outcomes. Part Two compares domestic and international judicial state-building in Afghanistan and Timor-Leste. The case studies draw on relevant primary and secondary sources as well as extensive fieldwork, and also on insights from overseeing judicial state-building initiatives in both countries. Important similarities make the comparison worthwhile. Afghanistan and Timor-Leste saw their infrastructures devastated by conflict and established new regimes in the early 2000s. Domestic and international state builders subsequently invested heavily in the state justice sector, working in an initially competitive legal pluralism environment where most disputes were settled by non-state mechanisms. However, the results in the two countries were very different. Despite periodic upheaval, judicial state builders in Timor-Leste made significant progress toward developing the rule of law by striving to build a democratic state with an effective state justice sector and a workable partnership with local non-state authorities. In contrast, the post-Taliban regime helped trigger a slide from competitive legal pluralism to combative legal pluralism owing to a lack of commitment to the rule of law, corrupt and unaccountable law and governance institutions, and a failure to engage key non-state actors. Ultimately the Afghan regime itself collapsed. While domestic actors and institutions have the most influence on judicial state-building outcomes, international actions can also have major consequences. Strategies that rely on large-scale spending or even the use of substantial military force in isolation are unlikely to be successful. The most promising approaches are culturally intelligible and constructively engage non-state justice networks of authority and legitimacy to collectively advance the judicial state-building process. While the case studies focus on post-conflict states, the theory and insights presented can help us understand and improve efforts to promote the rule of law as well as good governance and development more broadly in all legally pluralist settings.

4

From Competition to Cooperation

Building a New Legal Order in Timor-Leste

Introduction

The independent state of Timor-Leste was forged in fire. Throughout its history, East Timor featured extraordinarily high levels of competitive legal pluralism and, until recently, profound skepticism toward state institutions. The country endured over four hundred years of Portuguese colonial rule. Indonesia's occupation, which began in 1975, was even more brutal, leading to the death of roughly one-fifth of the population. Even after the overwhelming vote for independence in 1999, a further wave of Indonesian militia–backed violence engulfed the country. International peacekeepers were required to restore order. After independence, peacekeepers were again necessary to restore order in 2006 as domestic political contestation threatened to plunge the country into prolonged chaos. Ultimately, the 2006 crisis marked not the beginning but the end of large-scale unrest. Since then, democratic governance has persevered through serious incidents, including dual assassination attempts on the prime minister and the president as well as hotly contested elections in 2007, 2012, and beyond.

This chapter examines governance and judicial state-building in East Timor from Portuguese colonization to the final withdrawal of the U.N. mission at the end of 2012. The independence movement was committed to ideals of democracy and the rule of law. By and large, leaders of the independence movement, many of whom became the political elites of the new state, were able to translate ideals into policies, laws, and institutions. I argue that this success is due in no small part to the willingness of both the state and the non-state justice sector to work together since independence in 2002. This is a truly remarkable accomplishment given the state's limited capacity. It was

Contending Orders. Geoffrey Swenson, Oxford University Press. © Oxford University Press 2022.
DOI: 10.1093/oso/9780197530429.003.0004

made possible by a combination of historical factors and strategic choices. Historically, the relationship between state and non-state justice in Timor-Leste has been either competitive, under Portuguese rule, or combative, under Indonesia's occupation. In a break with decades and even centuries of earlier practice, the Democratic Republic of Timor-Leste (RDTL), with international support, established a legal landscape of cooperative legal pluralism between state and non-state justice within a decade. This transformation was accomplished by rooting local governing and judicial authority in competitive elections and establishing a functional jurisdictional divide between state and non-state justice.

In independent Timor-Leste, non-state justice actors supported state policy despite retaining meaningful autonomy and the fact that state-building efforts reduced, but did not eliminate, that autonomy. Judicial state-building efforts capitalized on the history, institutions, and ideals established during Portuguese occupation through to independence to pursue a popular, broad-based development agenda. This achievement, however, rests on a paradox. Portuguese rule did very little to develop East Timor or impose a uniform legal order. This neglect left indigenous social structures largely intact, particularly outside urban centers. At the same time, the Portuguese legacy proved a vital distinguishing feature from Indonesia, which was colonized by the Dutch, and allowed the resistance to articulate a compelling self-determination claim.

Summary of Arguments

In addition to challenging the conventional wisdom that the Timorese state is disconnected from society and that non-state justice is antagonistic toward the state, I argue that Timor-Leste offers a compelling case for theory-building of how a state, under the right circumstances, can undertake effective judicial state-building after conflict despite extensive legal pluralism. As highlighted in Chapter 3, the default backdrop for post-conflict state-building is competitive legal pluralism. However, the situation is inherently fluid. Timor-Leste shows it is possible for competitive legal pluralism to be transformed into cooperative legal pluralism. Timor-Leste was successful because (1) despite fierce political competition, all major parties remained committed to constructing a democratic state underpinned by the rule of law; (2) there were credible and sustained effort to develop institutions that promote democratic accountability, inclusiveness, and the rule of law; and (3) the state engaged and collaborated with key non-state actors through local suco elections as well as developing strategies to encourage suco chiefs to work with the state rather than against it. While the process was imperfect, Timorese state officials effectively mediated between the international community and local-level figures that contributed to the transformation of the competitive legal pluralist environment into a cooperative one.[1] Timor-Leste's development reflected its own history, for "the outcomes of the events during critical junctures are shaped by the weight of history, as economic and political institutions delineate what is politically feasible" (Acemoglu and Robinson 2012: 110). Still, these junctions, which occur in almost all post-conflict settings, are highly relevant to other state-building endeavors. As explored in Chapter 6, Afghan state leaders faced very similar decisions yet made very different choices.

Timor-Leste's success is not simply a function of good luck. The country faced serious challenges. Since independence, Timor-Leste has experienced a major political upheaval (in 2006) and numerous smaller but still significant bouts of political disorder. Yet, Timor-Leste has not descended into chaos but has enjoyed relative stability. Free and fair multiparty elections have become the norm. Non-state justice actors' support for the judicial state-building process bodes well not only for the state's legitimacy but also for its engagement strategy. It was not a lack of setbacks that ultimately determined the success of the judicial state-building project through to 2012, but rather how state institutions and leaders addressed

[1] State-building in Timor-Leste has been of significant scholarly interest, much of it decidedly critical. Critiques focus on three main areas. First, the U.N. intervention was misguided and misplaced and in later work often cast as substantially responsible for the 2006 crisis (Chopra 2000, 2002; Hohe 2002a; Butler 2012). Second, governance arrangements provided an insufficient role for non-state justice (Hohe 2003; West 2007). Third, the state, and particularly the state justice sector, lacked legitimacy because it was profoundly disconnected from Timorese society (Bowles and Chopra 2008: 272). This claim is often coupled with the idea that Timor-Leste's state institutions were largely a foreign imposition (Zaum 2007: 205–206; Richmond and Franks 2008). My research challenges all three of these claims.

While there were significant shortcomings, both during the U.N. interlude and since independence, these critiques tend to omit crucial developments. Few scholars even mention that the independence movement advocated for over two decades for a modern democratic state predicated on the rule of law. Those few who do acknowledge that fact tend to see it as a strategic calculation by the independence advocates rather than as a serious commitment (e.g., see Grenfell 2009). As this chapter highlights, in reality it was both.

these setbacks while remaining committed to rule of law ideals. This does not mean that state actors will uphold rule-of-law ideals on every occasion, which is an impossible standard, but whether a meaningful commitment to the idea exists.

Chapter Overview

This chapter consists of two parts. The first illustrates how state and non-state legal mechanisms have developed in tandem with, and frequently in opposition to, dominant modes of governance. It traces the relationship between state and non-state justice from the Portuguese colonial period, starting in the early seventeenth century, to the declaration by the dominant Fretilin party of an independent state in 1975 and the subsequent Indonesian occupation. It argues that Portuguese rule entrenched non-state justice in a manner still recognizable to this day, then details Indonesia's occupation and East Timorese resistance. Most notably, it shows legalism was a cornerstone both of Indonesian rule and of Timorese opposition to it. The resistance relied on non-state authorities and legal argumentation regarding self-determination, human rights, and a vision of an independent democratic state bound by the rule of law. Finally, this section details Timor-Leste's emergence from Indonesian occupation and how this process further empowered non-state authorities.

The second part examines judicial state-building since Indonesia's departure, as well as post-independence efforts to resolve tensions between justice systems from 1999 through 2012. It shows how the dual vision of a simultaneously traditional and fully modern state remained prominent during the U.N. interlude (1999–2002). Despite major challenges, the independent Timorese state garnered support from non-state judicial authorities for its efforts to institutionalize democracy and the rule of law. Competitive democratic elections have translated political legitimacy garnered from the independence struggle into state authority, while local elections rooted local authority in democratic choice rather than ancestry. Finally, this section explores how East Timor's leaders have established institutions and practices underpinned by a normative commitment to following the law, which has promoted progress toward the rule of law (Figure 4.1).

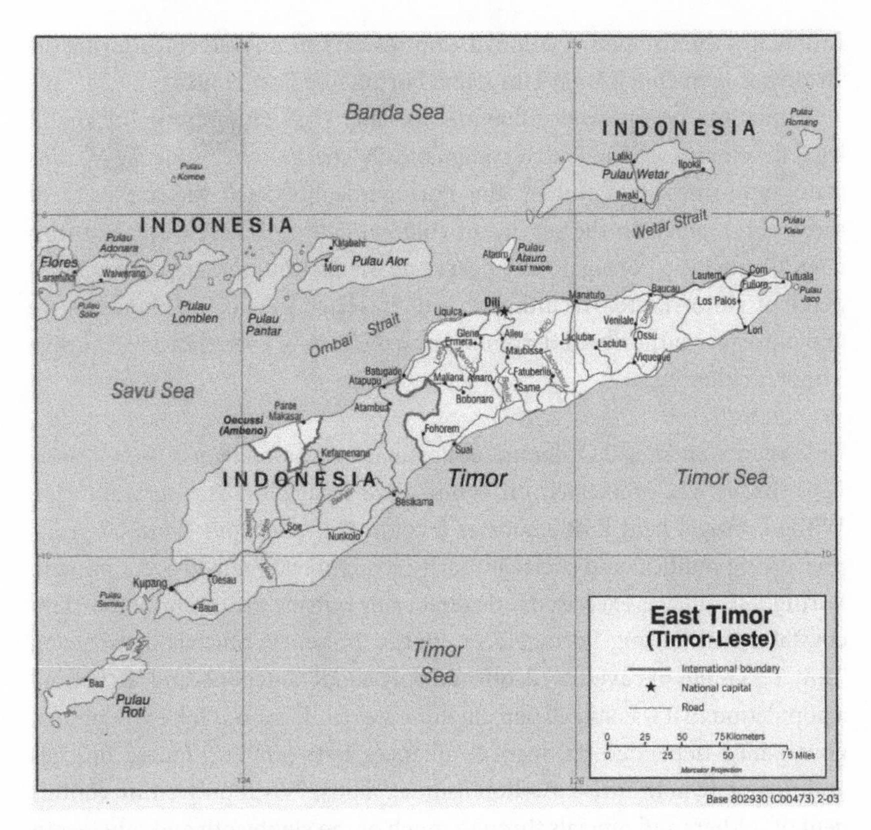

Figure 4.1 Map of East Timor
Credit: Central Intelligence Agency 2003, Library of Congress Map Collection, https://www.loc.gov/item/2003630243/

East Timor from Portuguese Colonialism to the Independence Referendum

Portuguese Colonialism

East Timor has long received colonial and international interest far disproportionate to its landmass, population, or economic clout. European colonial influence began with the arrival of the Portuguese in the early sixteenth century.[2] In the seventeenth century, the Dutch became a strong colonial rival.

[2] The terms "Timor-Leste" and "East Timor" are both used in this chapter; I generally use East Timor for the period prior to statehood and Timor-Leste for that after independence in 2002.

Eventually, the competing colonial administrations created an enduring division between Dutch West Timor and Portuguese East Timor.

Portuguese rule proved influential for later East Timorese political and legal development. The Dutch systematically checked the influence of non-state authorities. In contrast, the Portuguese bolstered the authority of those elites, albeit in the service of colonial rule (Jolliffe 1978). Centuries later, democratic Portugal would serve as the institutional model for independent Timor-Leste. The ultimate result has been an effective blend of state and non-state authority underpinned by a distinctive and relatively inclusive Timorese identity.

Governance and Legal Order under Portuguese Colonialism

Legal order was predicated on competitive legal pluralism for centuries. While Portugal held East Timor as a colony for over four hundred years, preexisting political and social authority structures remained predominant. Portuguese officials exercised little direct rule outside the capital and certain coastal areas. During Portugal's prolonged presence, officials constructed "only 12.5 miles of paved road, opened only about 50 schools and left behind a population that was still 80 percent illiterate" (Robinson 2009: 25). Colonial governance depended on tactical alliances between Portuguese officials and non-state authorities. As Robinson explains, "With only a tiny contingent of soldiers and officials through much of the eighteenth and nineteenth centuries, Portuguese authority was hardly overwhelming, and relied to a great extent on a network of informal alliances with friendly chiefs or liurai" (Robinson 2009: 23). Most issues were addressed locally, although certain matters, such as murder, could trigger state involvement. Indigenous East Timorese political structures featured three main administrative units: the village, consisting of several small hamlets, princedoms (sucos) composed of several villages, and the kingdom consisting of several sucos and headed by a *liurai* or ruler (Taylor 1991: 7). These core units remain essential to understanding both state and non-state authority through to the present day.

In the early twentieth century, Portuguese colonists faced a sustained rebellion. It was quashed in 1913 but sparked a transformation of the governance system:

> The position of the liurais was undercut by the abolition of their kingdoms. The colony was re-divided into administration units, broadly based on the suco (princedom). A measure of administrative power was given to the

> level below the kingdom level on the indigenous hierarchy. This enhanced the position of the *suco*, although their election as administrators was subject to Portuguese approval. (Taylor 1991: 11)

The emphasis on maintaining order through suco-level administration combined with a more direct approach by colonial administrators persisted until Portugal's withdrawal in the mid-1970s (Nixon 2012: 31–35). Liurais remained important, but suco chiefs emerged as another district source of non-state authority. In Portugal, the authoritarian Marcello Caetano regime collapsed in April 1974, which triggered a rapid, haphazard decolonization process. In East Timor, this meant legalization of political parties, significant internal conflict, and substantial international and domestic maneuvering over the territory's status. This turbulent period culminated in a declaration of independence by the majority Fretilin party on November 28, 1975.

Indonesian Occupation

Indonesia invaded East Timor on December 7, 1975, in violation of international law (Clark 1980). Indonesia subsequently undertook "one of the most intensive military occupations in history" (Moore 2001: 25). It resulted in the deaths of roughly 200,000 people in a population of fewer than a million (Nevins 2005: 26). Nevertheless, Timorese society remained remarkably resilient.

Legalism as a Foundation for Indonesian Occupation

In many ways, East Timor exemplified governance in President Suharto's Indonesia writ large. Suharto's regime claimed legitimacy based on its supposed support for the interests of justice over politics and a "return to the rule of law" compared to the previous regime (Hart 1987: 166). In reality, Suharto's Indonesia exemplified rule by law, wherein the regime uses law to exercise power, suppress opposition, and enhance its domestic and international legitimacy. Yet, the regime itself had no intention of being bound by law (Ginsburg and Moustafa 2008a: 4–6). The projection of normalcy and the rule of law formed a key component of Indonesia's strategy for dealing with the resistance (Robinson 2009: 77). Indonesia also justified its presence under international law. Indonesia claimed its motives were humanitarian and "it could not stand idly by while a small minority in Timor, represented by FRETILIN, ruthlessly imposed its ideas by force of arms" (Republic of Indonesia 1987: 53).

Indonesia argued that East Timor had already exercised self-determination through a representative People's Assembly, which had chosen to formally integrate into Indonesia in 1976 (Republic of Indonesia: 54–57).[3] The Suharto regime took pains to establish nominally elected local government structures in East Timor consistent with the practice in other Indonesian provinces (Dunn 1983: 301). Although suco chiefs were no longer trusted to maintain order, they were expected to actively support Indonesian rule. Even so, many suco chiefs played a double game as representatives of Indonesian authority at the local level and "clandestinely serv[ing] as the local CNRT [National Council of Timorese Resistance] representative" (Bowles and Chopra 2008: 274).

Courts were a cornerstone of Indonesia's attempts to solidify and legitimize its rule. They operated in almost all districts (Babo Soares 2003: 267) and marked a clear extension of state legal power relative to Portuguese colonial administration. Rather than extrajudicial killings, public trials became the norm for resistance leaders from 1983 onward (Commission for Reception 2005: Chapter 7.6). Public trials were also preferred for student protesters, who rose to forefront of the resistance after the Santa Cruz massacre in 1989. Formality, however, was not due process. Trials were not about justice. Trials sought to prevent the creation of new resistance martyrs, demonstrate the "far reaching authority of the Indonesian state" domestically, and present the veneer of due process and the rule of law internationally (Robinson 2009: 70–71).

Non-State Justice under Occupation

The Indonesian occupation reached deep into village-level structures and sought to uproot preexisting social, cultural, and political beliefs (Gunn 1997). Yet, most disputes continued to be settled through non-state mechanisms (Babo Soares 2003). In the conflict's initial phases, Fretilin also established a parallel justice system, which was outside the state but featured a significant degree of formalization (Commission for Reception, Truth, and Reconciliation 2005: see particularly Sec. 5, 12–14). After armed resistance collapsed, "higher order structures of leadership were

[3] This claim was vehemently denied by the resistance movement, which denounced the assembly as deeply flawed and unrepresentative because it was stacked with supporters of integration.

heavily modified or curtailed" by Indonesian authorities, but the indigenous "foundation of society" still regulated the social order (McWilliam 2005: 34). These networks mitigated collective-action problems related to "trust, secrecy, and organization" that were necessary to sustain the resistance and preexisting local institutions (Roll 2014: 486). Such structures enabled the non-state justice system to persevere despite the intense occupation.

Legalism as a Resistance Strategy

When armed resistance collapsed in the early 1980s and Xanana Gusmão assumed leadership, the resistance shifted its strategy. Initially, the resistance focused on armed struggle. From the mid-1980s onward, the East Timorese independence movement was notable for its highly formal, legalistic nature. Under Gusmão's leadership, the CNRT, a resistance umbrella group, recognized that winning a physical war was impossible. Support from the international community would be essential to get Indonesia to change its policy. Advocacy for nonviolent self-determination emphasized Indonesia's illegal occupation and systematic human-rights violations. Even those who had actively fought Indonesia did so cloaked in international legal claims. For instance, the Fretilin fighter Paulino Gama stressed that Indonesia's "invasion and occupation have been characterised by an open violation of international law" along with "a flagrant disregard for the relevant UN resolutions" (Gama 1995: 103). By the early 1990s, legal claims formed the core of the resistance movement (Pinto 2001).

Diplomatic efforts advanced East Timor's long-standing claim that Indonesia's "actions violated international law, specifically the norms regarding self-determination and aggression" and numerous U.N. resolutions dating back to the period shortly after the invasion (Clark 1980: 2).[4] In December 1975 the U.N. General Assembly passed Resolution 3485 stating that Indonesia should "withdraw without delay" (United Nations General Assembly, December 12, 1975). The U.N. Security Council followed suit by unanimously adopting Resolution 384 (1975), which demanded Indonesia's immediate withdrawal (United Nations Security Council, December 22, 1975). East Timor's top diplomat, Jose Ramos-Horta, sought to assuage Western concerns about an independent Timor by stressing that Fretilin did

[4] The U.N. General Assembly passed annual resolutions calling for Indonesia's withdrawal until 1982, and the U.N. Security Council passed a subsequent resolution in 1976 (Clark 2000: 80).

not hold Marxist ideals. He also aimed to discredit Indonesian legal claims and prove the right to East Timorese self-determination under international law and U.N. resolutions (Ramos-Horta 1987).

Gusmão, who likewise stressed self-determination and human rights, was captured in 1992. At his trial the following year, he decried the "illegal nature of the annexation, through force, of East Timor" as delegitimizing the proceedings (Amnesty International 1993a). Gusmão also drew a strong contrast to authoritarian Indonesia by underscoring that the resistance was "committ[ed] to building a free and democratic nation, based on respect for the freedoms of thought, association, and expression, as well as complete respect for Universal Human Rights" (cited in Webster 2003: 14). Gusmão also routinely emphasized that an independent East Timor would prioritize democracy and the rule of law, underpinned by "an independent judiciary" (cited in Strohmeyer 2000: 259).

Legal Activism

With approval from the central leadership, East Timorese students in Indonesia engaged in justice-based advocacy.[5] Student networks, like the broader resistance, publicized human-rights issues, formed linkages with sympathetic organizations, activists, and journalists, relayed information about activities within East Timor, and engaged in public advocacy (Braithwaite, Charlesworth et al. 2012: 67). Self-determination claims likewise formed the centerpiece of East Timorese student's asylum petitions at Western embassies in Jakarta (Amnesty International 1993b).

Timorese self-determination efforts were bolstered by solidarity movements and international NGOs worldwide that couched their arguments in the rhetoric of human rights and international law. Amnesty International, for example, produced reports throughout the 1980s and 1990s that helped keep international attention on East Timor (for example, see Amnesty International 1985, 1995). In sum, the independence movement correctly determined that international support was vital for any chance of success and that armed resistance was potentially counterproductive (Webster 2003). This legalistic approach, centered on self-determination, human rights, and international law, reached a far larger and more diverse audience.

[5] For a firsthand account, see Pinto and Jardine (1997).

Creation of East Timorese Nationalism

Indonesia's occupation, and the reaction it provoked, constructed and solidified an inclusive, broad-based East Timorese cultural identity deeply invested in a democratic Timorese state. That commitment was so deep that huge swaths of the population willingly risked their lives for it. The East Timorese identity, molded in the crucible of occupation, possessed certain key features. First, the occupation fostered a strong belief in a shared Portuguese heritage. While the Portuguese were not particularly benign or competent administrators, they were decidedly not Dutch or Indonesian. Moreover, Portugal's unwavering support for East Timorese self-determination since 1975 struck a positive chord. Second, Catholicism became an important cultural marker, one that provided a stark contrast to overwhelmingly Islamic Indonesia. Church membership rose from roughly one-third of the population in 1975 to over ninety percent in 1992 (Nixon 2012: 104–105). The church also adopted the widely spoken local language of Tetum for its liturgy, which helped further develop Tetum as the lingua franca and overcome extensive linguistic diversity (Taylor-Leech 2008). Finally, Indonesian authorities dramatically increased access to education (Jones 2000). While the quality was variable and the content highly ideological, an educated class committed to independence nevertheless emerged. Together these factors helped foster a broad-based, distinct East Timorese identity and, eventually, an independent state (Anderson 1993).

East Timor's complex reaction to occupation has prompted many commentators to argue not only that the occupation had a profoundly limited effect on non-state institutions, but at the same time also strengthened traditional authorities and effectively delegitimized the overarching concept of state authority in an independent Timor-Leste. Nixon, for instance, contends that "the nature of the Indonesian administration of East Timor merely acted to undermine the legitimacy of state authority and reinforce the prevailing 'traditional authority' ethic" (Nixon 2008: 183). Indonesia never succeeded in legitimizing its rule, in part because of local reliance on traditional justice mechanisms that represented the regime's fundamental illegitimacy to the local population. Yet it did not undermine the concept of a state or of state legal authority. The resistance wholeheartedly endorsed the concept of a modern Timorese state, provided it was the right kind of state led by the right kind of leaders.

Resistance to Indonesian occupation pushed East Timorese political elites to articulate a vision of an independent state that was compelling domestically and internationally. Despite their differences, the Fretilin leadership, Gusmão, and the broader political class all believed that East Timor needed development, which entailed harnessing the power of a modern state while still respecting local culture. It also meant a shift away from Marxist ideals to more mainstream political views. Even people in remote areas had experienced Indonesian development initiatives in areas such as education and infrastructure and increasingly accepted the notion that certain disputes were best handled outside of the community.[6] The desire for state-centered development even predates the occupation; it was Fretilin's aggressively statist rural modernization plans that underpinned their initial victory in the 1975 local elections (Taylor 1991: 32–35).

This combination of traditional values alongside a modern nation-state committed to the rule of law and human rights reflects how political elites envisioned the independent Timorese state, even if that vision was idealized. Society, particularly outside the urban centers, would be underpinned by "traditional" understandings of order, which are legitimate and effective for maintaining order outside the urban areas. State involvement would be generally reserved for matters outside everyday life—for instance, improving access to water, building roads, and other development projects. This meant that state authority could peacefully coexist with the non-state order.

During the conflict, however, these two visions—of a modern nation-state and a harmonious, custom-oriented society—were not accorded equal weight. The resistance movement was explicitly statist and developmental. It did not seek to return to an idealized, primordial East Timor. Non-state cultural practices helped sustain the domestic resistance and distinguish East Timor from Indonesian West Timor, but the resistance movement was already envisioning itself at the helm of a modern independent state long before statehood was on the horizon.[7]

[6] I traveled to nearly every district while I was working for the Asia Foundation. Remarkably, almost everyone I talked to stressed the need for development and the need for the state or international donors to provide it.

[7] Rhetorical commitment to democracy and the rule of law is sometimes nothing more than a strategy to secure independence (Huntington 1993: 47). Opposition to occupation or colonization can easily slide into authoritarianism after independence, as was the case following independence in Indonesia.

The Independence Referendum

At the beginning of 1997 few observers believed that East Timor was on the brink of major change. Then the Asian financial crisis struck, devastating Indonesia's economy and eventually leading to Suharto's resignation. B. J. Habibie became president in 1998; he immediately initiated democratization in Indonesia and undertook a new approach in East Timor. After extensive negotiations, Indonesia, Portugal (which retained colonial administration rights under international law), and the United Nations agreed to a referendum in August 1999. Timorese voters faced a choice between autonomy within Indonesia or independence. While the United Nations Mission in East Timor would conduct the referendum, "the agreement placed sole responsibility for maintaining law and order during the referendum" upon Indonesian security forces, who possessed a demonstrated propensity toward violence and had little interest in free and fair elections (Robinson 2009: 97).

The August 1999 Referendum and Its Aftermath

Before the referendum, the Indonesian military-backed militias undertook an intimidation campaign designed to secure a vote for autonomy (Nevins 2005). Still, a stunning 98.4 percent of eligible voters participated, with 78.5 percent supporting independence. Shortly thereafter, pro-integrationist militias unleashed a "systematic campaign of destruction and terror" (Smith and Dee 2003: 44). Jarat Chopra, who worked for the U.N. Transitional Authority in East Timor (UNTAET), surveyed the damage:

> More than three-quarters of the country's population of 890,000 were displaced. . . . Main cities as well as remote towns and villages were laid waste, and 70% of the physical infrastructure was gutted. Some areas were 95%-destroyed in street-by-street burnings more precise than smart-bombing. (Chopra 2000: 27)

In total, "between 5,000 and 6,000 people had been killed by paramilitaries since Habibie's announcement" (Taylor 1999: xxiv). The violence brought international condemnation and ultimately international peacekeepers in mid-September 1999. In October U.N. Resolution 1272 established UNTAET and vested it with territorial sovereignty during the transition to independence. During this chaotic time, non-state authorities still maintained order as the national political situation deteriorated from oppressive to near anarchical (Hohe and Nixon 2003: 2). While the United Nations asserted political sovereignty, non-state authorities maintained order locally, and state justice, to the extent that it was present at all, was clearly secondary.

Judicial State-Building and Legal Pluralism

From Independence to Statehood under UNTAET

UNTAET oversaw East Timor's transition from "neo-trusteeship" to full independence (Caplan 2005). UNTAET had "overall responsibility for the administration of East Timor" and was "empowered to exercise all legislative and executive authority, including the administration of justice" (UNSC 1999). As Tansey highlights, "The UN mission was charged with establishing the foundations of a future democratic government . . . but also the local capacity to use and maintain them" (Tansey 2009: 66–67). Despite the absence of antagonistic former combatants and widespread popular support, UNATET's task was daunting. The state institutions present under Indonesian rule had ceased to function or often even exist. Human-resource challenges were immense because Indonesia had prevented the development of a professional class.

Transition to Local Governance

The United Nations formally governed East Timor from October 1999 to May 2002. Sérgio Vieira de Mello, the U.N. Special Representative to the Secretary-General, possessed broad governance power. Vieira de Mello, however, soon faced pressure to include Timorese stakeholders in the government. UNTAET's mandate required it "to consult and cooperate closely with the East Timorese people" (United Nations Security Council 1999). The CNRT had been an effective umbrella group during the struggle for independence, but tensions between the CNRT leader Gusmão and Fretilin supporters, led by Mari Alkatiri, soon flared. This fundamental divide would ultimately shape Timorese politics for decades to come.

In December 1999 UNTAET established a fifteen-person National Consultation Council (NCC) to "ensure the participation of the East Timorese people in the decision-making process" (UNTAET 1999). The NCC, however, was criticized because it was dominated by Gusmão supporters (Tansey 2009: 71–72). "Timorization" began on April 5 with the commencement of the hiring process for seven thousand East Timorese civil servants, the hiring of local administers as deputy heads of district and centralized departments, and the establishment of the Civil Service Academy (Beauvais 2001: 1143). The NCC was replaced by the larger, exclusively Timorese National Council in July 2000 (UNTAET 2000a) as well as a joint U.N.–East Timorese Transitional Cabinet (UNTAET 2000b). While UNTAET still retained all formal authority, it had already delegated significant practical authority by the end of 2000.

Elections

The run up to the Constituent Assembly elections fundamentally transformed the political scene, as sixteen political parties quickly materialized. In June 2001 the CNRT dissolved, and Gusmão did not run for the Constituent Assembly. Over 90 percent of eligible voters participated in the Constituent Assembly elections on August 30, 2001 (Paris 2004: 219). The eighty-eight-member Constituent Assembly was empowered to draft a constitution, serve as a provisional legislature, and become the National Parliament after independence. Fretilin won 57.37 percent of the vote and fifty-five seats but lacked the "two-thirds majority it needed to be able to impose a constitution unilaterally" (Chesterman 2004: 231). The constitutional drafting process was completed by February 2002. Two months later, Gusmão was overwhelmingly elected president. The United Nations' rule officially ended on May 20, 2002, when East Timor became formally independent.

Even during this transitional period, the two major forces in Timorese politics, Fretilin- and Gusmão-affiliated groups, were forced to alternate power, compromise, and build coalitions. While the U.N. mission made mistakes, contemporary fears that either Fretilin or Gusmão had consolidated power to such an extent that the U.N. period had consigned the young country to chronic instability or a one-party state went unfulfilled (see, e.g., Hohe 2002b: 85).

The decision to disband the CNRT and allow the formation of political parties has been criticized as a missed opportunity to establish national unity (Hohe 2002b; Scambary 2009: 267). This argument, however, ignores the tensions within the CNRT. Criticism of the decision to dissolve the CNRT is a de facto proxy argument for Gusmão, rather than Fretilin, as the most powerful force during the constitutional drafting process. The CNRT provided an invaluable unified voice for independence, but post-independence politics were another matter entirely; there was never a unified CNRT policy agenda for governing. Political parties are essential for consolidating and sustaining democracy because they aggregate interests, allow voters to make meaningful policy choices, and sustain governments (Gunther and Diamond 2001: xiv; Lipset 2000). As will be discussed later, the process of institutionalizing the political party system has continued through the alternation of governments and the establishment of the country's first coalition government in 2007. The ability to change the government through free elections is a necessary, though not sufficient, condition for democratization and the development of inclusive rather than extractive political institutions. UNTAET established an institutional framework whereby the outcomes, be they good or bad, would reflect Timorese decisions.

Key Constitutional Structures

Despite thoughtful criticisms that UNTAET's approach to governance and institutional design laid the foundations for a one-party state (Chopra 2002; Hohe 2002a), a closer examination of the 2002 constitution shows it was a good-faith effort to create a modern democratic state that respects the rule of law and individual rights while recognizing traditional, non-state structures. It comports with the CNRT's earlier charter, which envisioned the inclusion of "traditional values within the new legal system and the new organisations of the State of East Timor as a state based on the rule of law" (CNRT 1998). The 2002 constitution combines an independently elected president with a prime minister who is elected and supported by a majority in the National Parliament, a unicameral body elected from a closed party list system with a three percent vote threshold to receive seats (RDTL 2002, 2006).[8] Legislative power rests primarily with the National Parliament, though the Council of Ministers, appointed by the prime minister, can also initiate certain legislation. The right to form a political party is also protected (RDTL 2002, Sec. 70).

The president cannot initiate legislation or promulgate regulations but can issue a veto. Vetoes can be overturned with a simple absolute-majority vote in the National Parliament. The president possesses some limited executive authority, most notably the power to pardon, but overall the role is relatively weak (Kingsbury 2013). While the prime minister is unquestionably the dominant actor, the constitutional arrangements provide for more and

[8] The constitution limited Parliament to between fifty-two and sixty-five members (section 93). The electoral law of 2006 permanently set the number of members at sixty-five (RDTL 2008a).

stronger checks on the prime minister's power than are commonly found in a "pure" parliamentary system. Thus, Timor-Leste can be classified as a "parliamentary-like semi-presidential regime," with a strong prime minister and a relatively weak president, an arrangement that tends to have a positive effect on democratization (Elgie 2005).

Constitutional Structure of the Legal System

Reflecting the broad appeal of the rule of law and human rights among Timor-Leste's political elite, the constitution declares Timor-Leste "a democratic, sovereign, independent and unitary State based on the rule of law, the will of the people and the respect for the dignity of the human person" (Sec. 1.1). State activities must be consistent with the constitution and the law (Secs. 2.2 and 62).

Timor-Leste is a civil-law country closely modeled on Portugal, with protections for judicial independence, due process, individual rights, and access to courts.[9] The constitution provides for a Supreme Court, a Court of Appeal, district courts, and specialized courts (Sec. 123). The Superior Council for the Judiciary oversees "the management and discipline of the courts" along with the appointment, assignment, and promotion of judicial personnel (Sec. 128.1).

Constitutional clauses alone do not guarantee independence. Even in consolidated democracies, courts may succumb to political pressure (Rosenberg 1992), and the problem is far more pronounced in places lacking a tradition of judicial independence (Vyas 1992). While Timor-Leste's institutional arrangements may ultimately prove unable to establish or sustain the rule of law, the constitution clearly envisions an independent judiciary and outlines a credible framework for establishing one.

The constitution explicitly recognizes non-state justice and its ideological foundations. Under section 2.4, "The State shall recognise and value the norms and customs of East Timor that are not contrary to the Constitution and to any legislation dealing specifically with customary law." How the state conceptualizes and deals with non-state legal authority will be discussed further in subsequent chapters. Still, it is worth noting here that based on my discussions with practicing lawyers, judges, officials at the Ministry of Justice and the Office of the President during my work in Timor-Leste from 2010 to

[9] See Merryman and Pérez-Perdomo (2007) for a succinct overview of civil law systems.

2012 and during subsequent field research, non-state justice was taken very seriously as a legitimate dispute resolution system.

While the constitution and the state structures it established have been faulted for failing to reflect Timorese values (Zaum 2007: 205–206; Jones 2010), this critique fails to appreciate the document's broader political context. Whatever its strengths and weaknesses, the constitution reflects the independence movement's aspirations. The constitutional structures are overtly modeled on those of democratic Portugal—a persistent opponent of Indonesia's occupation. Timor-Leste's constitution envisions a modern state underpinned by the rule of law, multiparty democracy, and fundamental rights while recognizing the importance of norms and customs and allowing non-state dispute resolution. As such, the constitution represents a good-faith effort to institutionalize the independence movement's vision.

Establishment of the State Legal System

The legal sector did not escape the devastation surrounding the referendum. U.N. Secretary-General Kofi Annan reported that "local institutions, including the court system, have for all practical purposes ceased to function" and that "members of the legal profession hav[e] left the territory" (Annan 1999b: para. 33). East Timorese legal professionals could not serve as prosecutors or judges under Indonesian rule (Chesterman 2004: 170). Hansjoerg Strohmeyer, who oversaw much of the United Nations' initial legal sector efforts, contends that "all court equipment, furniture, registers, records and archives . . . ,. . . . law books, cases files, and other legal resources were lost or burned" (Strohmeyer 2001a: 50). While the process was by no means perfect, even critics such as McAuliffe have recognized that "the UN was successful in improving the functionality of the courts" through a process of "defining the applicable law, appointing judges, and creating the District Court System" (McAuliffe 2011: 115).

UNTAET pursued "Timorization" of the justice sector aggressively from November 1999. The United Nations distributed flyers by plane asking anyone with legal experience to get in touch. Although approximately sixty people responded, "only a few of these jurists had any practical legal experience," and none had ever worked as a prosecutor or a judge (Strohmeyer 2001a: 54). Nevertheless, two prosecutors and eight judges were appointed by early January 2000 (Strohmeyer 2001a: 54). Despite concerns about their qualifications, U.N. administrators continued to appoint Timorese judges, prosecutors, and public defenders throughout the UNTAET period. Yet, there still were still too few legal professionals, and they were often poorly trained. By March 2000 it had become clear that domestic personnel alone would be insufficient, and international court actors were hired (UNTAET 2000c, d). As Howard notes, "There was simply not enough time before independence in May 2002 to train the numbers of people required to run an effective, efficient, and politically independent legal system" (Howard 2008: 284).

Overview of Non-State Justice

While always a powerful force, the transition from independence further strengthened the non-state justice system's legitimacy and reach (Nixon 2012: 117–119). Non-state justice practices varied, but certain core facets were commonly observed in places operating along traditional lines:

> The process of applying indigenous law starts with a report of the issue to the village or hamlet chiefs (depending on the level on which the conflict or the crime occurred) by the heads of the families involved in the conflict, or the family of the victim. The "helper" takes note and reports to the "local legal authorities," such as the *lian nain*. The *lian nain* know the history and are in contact with the ancestors. They come from specific families that are the "owner of the words." They know the rules the ancestors have set and, therefore, they have the competence to speak the law. (Hohe 2003: 343)[10]

A mutually agreeable time for the dispute resolution process would be set shortly thereafter, often by the suco chief. The process would involve a structured process of discussion, testimony, and negotiations before reaching a final decision.

Centered on small, tightly knit communities, non-state justice emphasizes compensation and reconciliation rather than punishment, a focus that likely predates even Portuguese colonial rule (Berlie 2000). Compensation can involve payment of money, livestock and other goods, services, and land (Nixon 2012: 175–182). Once appropriate compensation has been determined, reconciliation seeks to restore communal harmony (Babo-Soares

[10] For more detailed discussions of the practice of and beliefs behind non-state community-based dispute resolution processes, see, e.g., Hohe (2003); Babo-Soares (2004); Meitzner Yoder (2007); Trindade and Castro (2007); Cummins (2010); and Nixon (2012).

2004: 23). Thus, even if a judgment has been rendered in state court, the matter may still be unresolved locally.

For dispute resolution to be acceptable, judgments must have both authority and legitimacy. While most disputes are quickly resolved within the suco, appeal mechanisms exist (Ospina and Hohe 2001: 114–120; Nixon 2012: 175). This process is often cast as mediation. In reality, it is more akin to arbitration, for the social pressure to accept a decision can be intense. Historically, it could even involve physical violence (Tilman 2012: 198). Certain disputes, however, were recognized as outside the purview of the non-state system, most often defined as serious crimes that involve the shedding of blood.

While non-state justice systems were still commonplace, particularly in remote areas, their reach was far from uniform. In Dili, for instance, traditional understandings of authority have broken down owing to migration and broader social and economic changes (Wassel 2014). This trend is particularly pronounced amongst younger Timorese (Scambary 2006). Tilman provides a useful schematic for understanding non-state justice in contemporary Timor-Leste:

> The first category considers those *suku* in which the *liurai* no longer have any real power, but the *liurai*'s *uma lisan* continues to be strong. The second category covers those "new suku" that were formed during Indonesian occupation, often comprising of people from different areas, and (as a result) in which both the *liurai*'s governing power and the *liurai*'s *uma lisan* are not present. In the third category, there are *suku* that are not new, but where for various reasons the influence of the *liurai* and the *liurai*'s *uma lisan* appear to have died out. In the final category, there are *suku* in which the power of the *liurai* and their *uma lisan* remains strong. (Tilman 2012: 1999)

The forms and authority of non-state justice differ, but, as will be discussed later in this chapter, Timor-Leste has moved from competitive to cooperative legal pluralism. All manifestations of non-state justice now depend upon democratic electoral results and support the modern state-building process.

Non-State Justice under UNTAET

Preexisting non-state structures flourished under UNTAET's administration. The vibrancy and authority of non-state justice contrasted starkly with the low capacity, legacy of illegitimacy, and administrative confusion of the state courts. As national level political issues came to be addressed and the nascent state justice system was established, non-state legal structures were not prioritized (Bowles and Chopra 2008: 272).[11] UNTAET largely ignored non-state justice entities from a regulatory standpoint. U.N. officers even sent cases to the non-state system. For instance, a U.N. civilian police officer who worked in Baucau and Oecusse noted that "we had to 'offload' the majority of matters to a traditional resolution because we did not have the time or manpower to deal with all offenses in a formal way" (quoted in Nixon 2012: 185).

Independent Timor-Leste (2002–2012)

Independent Timor-Leste faced profound developmental and political challenges. The state has accomplished far more progress in democratic state-building and the rule of law promotion than most critical observers concede, and it has benefited from leaders and citizens generally committed to democracy and the rule of law. Timor-Leste made significant progress toward a variant of democracy and the rule of law that blends modern norms of justice with more traditional, local approaches to dispute resolution. Democratic governance has persevered through major crises, thanks to regular free and fair democratic elections that have translated the independence movement's legitimacy into authority for the nascent state, increasing the reach and capacity of the state court system, and most impressively, securing the cooperation of the non-state justice system for its state-building agenda.

The 2006 Crisis

The 2006 crisis was the most disruptive event in independent Timor-Leste's short history. It resulted in thirty-six deaths, the displacement of 150,000 people, and the destruction of over 1600 homes (U.N. Independent Special Commission of Inquiry for Timor-Leste 2006: 42). The crisis also led to renewed international intervention and the collapse of Fretilin Prime Minister Mari Alkatari's democratically elected government.

[11] The most significant exception was the Commission for Reception, Truth and Reconciliation (CAVR) established by UNTAET to examine past human rights violations. The CAVR incorporated indigenous structures, though this was largely due to the initiative of the commission itself (Nixon 2012: 189–192).

In January 2006 a group of soldiers, often dubbed the petitioners, complained to high-ranking officials that soldiers from western Timor-Leste were facing discrimination by officers from the east. Over the next few months the petitioners' ranks grew, and tensions increased in April 2006 after roughly one-third of the army was dismissed for refusing to return to their barracks. The situation then deteriorated further:

> An additional group of military and civilian police, led by Major Alfredo Reinado, abandoned their posts in protest of the military's April 28 deployment. This group ambushed soldiers and police three weeks later on May 23, killing five. Over the next two days, violence exploded in Dili, with petitioners and elements of the police engaging in armed battle with the military. Individuals attacked the home of Army chief Brigadier General Taur Matan Ruak, killing one civilian, and set fire to Minister of the Interior Rogerio Lobato's house, killing six. The deadliest incident involved soldiers firing on unarmed police under U.N. escort, slaying nine. Large-scale clashes between "eastern" and "western" gangs also took place. (Nevins 2007: 165)

The crisis severely tested Timor-Leste's emerging institutions. Some observers even claimed that East Timor was becoming a failed state.[12] Yet, the state decidedly did not fail. While international troops provided security, the political impasse at the heart of the crisis was resolved through constitutional means. Moreover, as non-state judicial authorities largely maintained order within their respective localities, the crisis was primarily a Dili-based event.

U.N. troops returned to East Timor to restore order in July 2006. The deeper reasons behind why the crisis occurred and how it unfolded are complex and contested (Scambary 2009; Gledhill 2014), but the intense political rivalry between then–Prime Minister Alkatiri and President Gusmão was crucial. Although Alkatiri resigned and was replaced in a manner consistent with the constitution, the process that propelled his resignation raised significant questions regarding the viability of parliamentary governance and the relative impunity enjoyed by perpetrators (U.N. Independent Special Commission of Inquiry for Timor-Leste 2006: 2).

[12] For a critical examination of the "state failure" discourse with regard to Timor-Leste, see Cotton (2007).

Other post-crisis developments were more promising. The free and fair elections the following year resolved elite competition without violence. In 2008 Timor-Leste handled the political upheaval related to the dual assassination attempts on the president and prime minister in accordance with the constitution and without significant violence, which in turn improved the security situation (Tansey 2009: 106). Democratization and social upheaval are by no means mutually exclusive. Conflict can obviously be destructive to democratic state-building, but movement toward further democratization "generally occurs in the face of significant conflict and the possible threat of revolution" (Acemoglu and Robinson 2005: 29). The 2006 crisis strained the Timorese state, but also created a distaste among political elites for open, potentially violent confrontation that could easily undercut progress toward the rule of law. Since the crisis, key political actors have shown consistently decreasing interest in extrajudicial attempts to undermine the state.

National Electoral Competition

Electoral competition is both a potential trigger for and a solution to violence (Sisk 2009). Tensions were high after the crisis. Nonetheless, free and fair presidential and parliamentary elections were held in 2007. All major political parties accepted the results. Jose Ramos-Horta was elected president as an independent, but with Gusmão's backing. In the parliamentary poll, Fretilin again secured the largest percentage of seats (twenty-one seats and 29 percent of the actual vote) but not a majority. Gusmão's new party, the National Congress for Timorese Reconstruction (also dubbed CNRT), won eighteen seats with 24.1 percent of the vote.[13] Over Fretilin's strenuous objections, Gusmão formed a coalition government with smaller parties. At least part of Fretilin leadership initially stoked violence in the wake of the poll, but within months Fretilin's entire leadership respected the result and remained committed to democratic competition rather than seeking to undermine the government. The 2007 elections marked Timor-Leste's first democratic transfer of power and the emergence of a strong parliamentary opposition.

Movement toward democratic consolidation continued with the 2012 national elections, which were procedurally successful, substantively open, and highly competitive. The elections were overseen by national election bodies

[13] References to CNRT from this point forward refer to the political party rather than the independence front.

without significant external support. Taur Matan Ruak was elected president with Gusmão's backing. The CNRT secured just over 36 percent of the vote, while Fretilin earned just below 30 percent. Only four parties cleared the 3 percent vote threshold for seats. Consequently, the CNRT received nearly 45 percent of seats but still needed to form a coalition for a parliamentary majority.

While democratic consolidation remains incomplete, these first three national elections mark positive steps toward democratic consolidation and the institutionalization of a party system. There was a peaceful change of power, a successful nonviolent resolution of a major constitutional debate after the 2007 polls, and two distinct governing coalitions with the opposition increasingly eschewing violence and respecting poll results. These accomplishments can by no means be taken for granted. In Afghanistan, by contrast, elections have generally been marred by violence, large-scale fraud, and other irregularities and have shaken its people's faith in the political order rather than bolstered it.

The major political parties effectively draw on the independence struggle for legitimacy and electoral advantage. Since the Constituent Assembly elections in 2001, Fretilin has emphasized its link to the struggle for independence and the party's role in the formation of an East Timorese national identity (Hohe 2002b). The flag of Timor-Leste explicitly incorporates that of Fretilin. When Gusmão formed a new political party, it "drew heavily on images of independence and modernity with Xanana himself the prominent public face of all CNRT campaigning" (McWilliam and Bexley 2008: 69). The party name, National Congress for Timorese Reconstruction, sought to remind voters of Gusmão's leadership of the National Council of Timorese Resistance during the independence struggle. Gusmão's two organizations even share a common acronym, CNRT. In some ways, electoral competition reflects competition between "FRETILIN and former CNRT figures over the symbolic 'ownership' of the resistance, and its narrative of national liberation" (Leach 2006: 233). The third largest party, the Democratic Party, highlighted its role in the clandestine movement, and smaller parties consistently stressed their own connections to the independence struggle (McWilliam and Bexley 2008: 69–70).

The major political parties largely succeeded in linking themselves to the independence movement. This association bolsters the state's ability to govern because it is not an abstract entity but rather a known collection of groups and individuals that enjoy widespread legitimacy. The divided independence movement has led to sporadic clashes. At the same time, divisions have helped lay the foundations for multiparty democracy, since no one group has sole claim to the independence struggle's legacy.

The Transformation of Non-State Justice

Law and governance in Timor-Leste can accurately be described as a hybrid model, where non-state mechanisms play an important role in governance (Meitzner Yoder 2007; Brown and Gusmao 2009; Richmond 2011). While suco authorities perform some statelike functions, state officials do not consider them state actors (RDTL Ministry of State Administration and the Asia Foundation 2013). Likewise, suco officials do not view themselves as state officials and actively opposed formal the integration of non-state justice into the state system (RDTL Ministry of State Administration and the Asia Foundation 2013). In 2009 the Court of Appeal rejected a claim that the 2009 Community Authorities Law was unconstitutional because the elections were nonpartisan. The court's judgment reaffirmed that suco councils predated the state itself and thus occupied a sort of liminal position as "traditional organizational structures" distinct from local government or an NGO (RDTL Court of Appeal 2009).[14]

Suco councils are outside the formal state structure but are still legitimately subject to state regulation as non-state entities. Building on the precedent set during Indonesian times, at the subnational level elections for suco chiefs were codified under state legislation in 2004, with additional reforms in 2009 (RDTL 2004a, b, 2009a).[15] Suco elections were held in 2004–2005 for 442 councils, and a subsequent set of elections was undertaken in October 2009 (Maia, Cullen et al. 2012: 10).[16]

[14] The positionality of suco councils as non-state entities was later reinforced by the 2016 Law on Sucos (RDTL 2016).

[15] Sucos averaged roughly two thousand to three thousand people, but they could vary widely. Comoro, in Dili district, was the largest, at 65,404, while there were a mere fifty-four people in Caicau, in the Baucau district (RDTL National Statistics Directorate and United Nations Population Fund 2011).

[16] Under the changes passed in 2009, suco elections became officially nonpartisan. and suco chiefs and members of the suco council are now elected as a slate rather than individually (RDTL 2009a). These changes were made largely in response to the view that the previous election system had "interpolated the polarization that characterized elite politics into the grassroots, raising levels of insecurity and undermining social cohesion" (Brown and Gusmao 2009: 67). There have been some minor changes since 2012, most notably the 2016 legislation on sucos, but the framework remains fundamentally the same (RDTL 2016). The successful 2016 suco elections were held under this framework.

State Regulation of the Non-State Law and Governance Actors

The suco election process marks the state's most significant regulation of the non-state justice system. Dispute resolution ranks as suco chiefs' most important responsibility. Under the 2009 Community Authorities Law, suco chiefs can resolve "minor disputes involving two or more of the suco's villages" (RDTL 2009a: Art. 11.2(b)). With independence, these positions became elected "through secret, free, equal and direct ballot of community members" (RDTL 2004b). The election itself is regulated and validated by the state—a system that has provoked little protest and found widespread agreement. As the inaugural regulation explicitly states, "The election of suco chiefs and suco councils is of paramount importance to legitimise community authority and develop the basic structures of such authority" (RDTL 2004b). Older, well-established men traditionally dominated suco councils. The regulation thus sought to promote gender equality and broaden societal representation on the councils through requirements that there be two female representatives on the council and two youth representatives (one male and one female) (RDTL 2009a: Article 5).[17]

The Community Authorities Law was largely procedural and jurisdictional. However, its goal was transformational: it rooted the legitimacy of non-state judicial actors in modern democratic ideas, most notably through recurring, state-administered democratic elections and a workable jurisdictional divide. This concern became particularly pressing after the 2006 crisis. As Peira and Lete Koten observe, "Ideally. . . elections for local leaders transforms the relationship between community leaders and their constituents from *authority over* the community, to *representative of* the community" (Pereira and Lete Koten 2012: 223, italics in original). It is striking how quickly and comprehensively state authorities established democratic elections as the primary source of legitimacy, and how this foundation of legitimacy has been internalized by the suco chiefs (RDTL Ministry of State Administration and the Asia Foundation 2013). While many local leaders hailed from traditionally elite families, elections brought "a substantial portion of newcomers to the positions of village and hamlet chiefs" (Graydon 2005: 69; Brown, M. A. 2012). Well-established

[17] The results, however, have been mixed, as the "women's representatives, youth members, and *xefe aldeia* [are] often viewed as passive and unqualified for membership" (RDTL Ministry of State Administration and the Asia Foundation 2013: 22). Still, women have occasionally been elected suco chiefs.

and respected figures still possess electoral advantages. As Tilman succinctly notes, "Democracy does not deny opportunities for *liurai* to obtain positions of political power through a political party or as an independent candidate" (Tilman 2012: 197). Most notable is not that *liurai* or other non-state leaders seek political power through elections, but that even traditional leaders now require democratic ratification. The law even empowers the suco council to appoint the *lian nian* for his or her term (RDTL 2009a: Art. 5.3).

The state's transformation of the relationship between state and non-state judicial authority, and governance authority more generally, has been nothing short of revolutionary. After all, "the major struggles in many societies, especially in new states are over the right and ability to make the countless rules that guide people's behavior" (Migdal 1988: 31). These tensions can be particularly intense in a new state such as Timor-Leste, where the non-state authorities were entrenched and enjoyed independent legitimacy. Non-state actors formed the heart of local resistance to Indonesian rule, and the Indonesian state itself lacked legitimacy. Thus, non-state actors enjoyed more or less plenary power during the transition to independence.

By 2012 non-state authorities formed the backbone of the Timorese state outside the urban areas. While state capacity was limited, it enjoyed real legitimacy. Call, for instance, cites Timor-Leste as a paradigmatic example of the class of "legitimate states with low capacity but high legitimacy" (Call 2008b: 1496). The state's legitimacy helped transform non-state judicial actors, who had enjoyed both the independent authority and legitimacy necessary to seriously challenge the state, into almost de facto state actors, through elections and financial support, and by offering a vision of the state that commands widespread support. The state-administered suco election system forms the backbone of state influence at the local level, despite suco council members' ability to disregard directives from state authorities or act outside their legislative mandate (RDTL Ministry of State Administration and the Asia Foundation 2013). The most pressing debates do not center on whether democratic elections are appropriate for allocating suco authority, but on whether suco authorities should be fully incorporated into existing formal state structures and whether the state's expectations of suco authorities are realistic (RDTL Ministry of State Administration and the Asia Foundation 2013).

From Competitive to Cooperative Legal Pluralism

Historically, the relationship between state and non-state justice in Timor-Leste has been overwhelming competitive: tactical engagement coexisted alongside deep skepticism. One of the nascent state's greatest achievements has been facilitating a transformation into cooperative legal pluralism whereby state and non-state authorities worked together to promote development and access to justice (Wassel 2014). The state handles major issues, while most civil matters and petty crimes are resolved locally.

Non-state authorities continued to resolve most disputes. Police officers still referred cases to suco councils. Since independence, non-state authorities have enjoyed consistently higher approval than their state counterparts (Asia Foundation 2004; Everett 2009; Marx 2013). Nevertheless, non-state authorities viewed the state and the state-building endeavor as legitimate and supported constructive engagement because state leaders were also recognized as key actors in the struggle for independence.

During this period, more direct state regulation of non-state justice was never enacted, despite significant interest from the MOJ, the UNDP, and other bodies. Consequently, the primary regulation was the non-state justice sector Community Leadership Law. It stipulates that community leadership should structure "the community's participation in the solving of its problems" (RDTL 2009a: Art. 2) as well as resolve local disputes (RDTL 2009a: Art. 11.2(b)). In other words, the law already makes a formal distinction between minor crimes, which can be handled at the suco level, and more serious ones, which demand state action.

Existing regulation, while minimal, worked well, because it reflected prevailing social views regarding how disputes should be handled. Moreover, the state resisted the temptation to overregulate a sector that enjoys widespread legitimacy. While the state could easily impose its will on an individual community, it would be hard to consistently assert control over all localities simultaneously. In addition to the position's intrinsic status, the state offers enough inducement to make the post worthwhile, but not excessive. The suco chief received US$100 per month, with $20 earmarked for transportation and $15 for administration (Maia, Cullen et al. 2012: 9). Suco chiefs were also given an office and a Mega Pro motorbike. While suco chiefs routinely cited the need for increased compensation, these incentives are by no means trivial, for "half the population is estimated to live below the poverty line and around two-thirds is considered food insecure" (Coordination

Team of the U.N. System High-Level Task Force on the Global Food Security Crisis 2010: 2).

The suco reforms enacted by the 2009 legislation also subtly changed the suco chiefs' role by more tightly linking it with the regime's overarching development project. This includes working with relevant state authorities, conducting planning, program monitoring, creating an annual development plan, and submitting an annual report (RDTL 2009a: Arts. 11–12). These legislative responsibilities were paired with major state initiatives putting funds into local communities for development projects.[18] Suco chiefs were now expected to secure state funds to support development. On the one hand, this is unsurprising. Politicians are generally expected to access state resources for their constituents, particularly when directly elected (Lancaster 1986). Not only are suco chiefs active state-building agents—their constituents demand it. But the results are surprising if the suco leaders merely wanted to maximize independence from the state or maintain preexisting patterns of behavior. After all, sucos acquiesced in state-administered leadership elections, and these processes enjoy widespread legitimacy. Some suco chiefs certainly benefit from their ancestry, but even their continued tenure hinges on electoral success.

Tensions in Non-State Administration

Non-state authorities are not formally state officials but perform important functions for the state. Therefore, suco actors raise significant principal-agent incentive issues (Laffont and Martimort 2002). Delegation involves very high agency costs because the state's control mechanisms are quite weak (Jensen and Meckling 1976). The Timorese state lacks the capacity for the systematic monitoring and evaluation often seen as essential for mitigating these issues. While principal-agent problems cannot be eliminated, they can be mitigated. The state has made significant strides in aligning its overarching interests with those of the suco councils. As Fukuyama explains, "Norms can embed the interests of the principal," allowing the agent to significantly reduce the problem (Fukuyama 2004: 65). He cites the example of teachers going beyond their contractual requirements to provide better education for their students.

[18] These initiatives included the Ministry for State Administration and Territorial Management's Local Development Program, started in 2004; the Decentralized Development Program, started in 2010 by the prime minister's office; and the National Program for Village Development, launched in January 2012, to cover all sucos. Suco chiefs played a major role in determining local community priorities in these initiatives and state development projects more generally (Cummins and Maia 2012).

The same logic holds true for local leaders promoting their community's interests. Suco council members, and particularly suco chiefs, are important leaders and subject to sustained electoral scrutiny. Suco residents can punish bad leaders and reward good leaders through democratic accountability mechanisms. This dynamic makes state monitoring less crucial; suco chiefs largely accept the state's program and face significant electoral pressures that tend to align with the state's development initiatives.

That alignment was not perfect, however. Cases of domestic violence were a frequent area of tension between state and non-state legal orders. As Hohe notes, "Under the Indonesian law as well as in the traditional system, for example, domestic violence was a purely private matter" (Hohe 2003: 345). Since independence, political elites have embraced progressive reform. Domestic violence is unequivocally a public crime under the Penal Code (RDTL 2009b: Arts. 146, 154). The Law against Domestic Violence (RDTL 2010b) went further, stipulating that victims cannot drop their claims once legal proceedings have been initiated. Suco chiefs were encouraged to "promote the creation of mechanisms for preventing domestic violence" as well as protecting victims and "punishing the aggressor, in such a way as to eliminate the occurrence of said situation in the community" (RDTL 2009a: Art. 11.2(c–d)).

While domestic policymakers and the international community prioritized promoting gender equity and protecting women's rights, this paradigm shift from private matter to public crime went against much established state and non-state practice (Corcoran-Nantes 2009; Cummins and Guterres 2012). Despite serious investment in the state system, the non-state justice system continued to resolve most domestic-violence cases (Wigglesworth 2013).[19] How suco leaders address domestic violence remained largely discretionary in practice, which gives non-state authorities an effective veto over state law in their jurisdiction. Moreover, women were often unable to access state courts, and state officials frequently ignored their claims despite the law's written requirements (Cummins and Guterres 2012). Thus, non-state authorities' willingness to be directed by the state in most matters reflected the state's legitimacy and persuasive authority rather than its ability to uniformly impose its will.

[19] Cases stayed in the non-state system for various reasons, including a lack of interest by state authorities, the prolonged judicial process, little support for women outside the courts, little education about the relevant law, and "the inability to access important rights within civil law if the situation results in separation or divorce" (Cummins and Guterres 2012: 4).

The State Legal System

Building a state justice system faced an array of difficulties arising from minimal human resources, limited judicial infrastructure, and the public impression inherited from Indonesian rule that courts were instruments of state power rather than neutral arbitrators. Certain policy choices have further complicated this task. The legal system retained a myopic focus on Portuguese—which most people, including those with university educations, cannot understand—as the preferred legal language and the importation of Portuguese laws that had little relevance to the local context. At the same time, the legal sector was simultaneously over- and underregulated. Strict protocols regulated the practice of law, but there was no regulation of the private universities that churned out large number of law graduates with very dim employment prospects. Still, there were important signs of progress.

The Default Legal Framework

When international forces departed Timor-Leste at the end of 2012, the legal framework had been largely localized. Indonesian law remained the default law at independence prior to the passage of new legislation, as had been the case under UNTAET. By 2013 basic laws were in place in most areas necessary for a modern state and economy (UNMIT 2011). Legislation specific to the judiciary included a criminal and civil code, establishment of prosecutor and public defender services, regulations for private lawyers, and a host of other relevant regulations. While legislative gaps remain, law now reflected domestic preferences.

Development of State Justice Institutions

Even before the 2006 crisis, the justice sector was "the weakest branch of Timor-Leste's governance infrastructure" (World Bank 2006: 19). Between independence in 2002 and the 2006 crisis, procedural due-process concerns were endemic, as were case backlogs and spotty opening hours (West 2007: 336–338). The UNDP observed, for instance, that "27% of the total inmate population of approximately 292 are held without valid committal warrant" (UNDP 2002: 2).

The judiciary's rapid "Timorization" during UNTAET won praise from commentators critical of what they perceived to be the slow process of bringing Timorese elites into government more generally (Beauvais 2001). Yet, this strategy also raised difficulties, because local judicial actors were inexperienced and needed extensive training. In 2005 all UNTAET-appointed Timorese court actors were evaluated. Consequently, "the East Timorese government disqualified *all* Timorese judges, prosecutors, and public defenders in 2005" (Jensen 2008: 133, italics in original). This left the justice sector almost entirely dependent on international staff and raised doubts about the legitimacy of previous decisions. Yet, even this difficult time saw progress, with the appointment of the President of the Court of Appeal and the Superior Council of the Judiciary in 2003.[20] Furthermore, the increased presence of international court officials mitigated the impact of the wholesale dismissal of Timorese judicial actors and significantly decreased the case backlogs (World Bank 2006: 19–20).

In 2004 the Legal Training Centre (LTC) was established to oversee the training and professional certification of all judges, prosecutors, and public defenders (RDTL 2004c). After the passage of the Private Lawyers Law in 2008, the LTC began training private-sector lawyers (RDTL 2008). Since 2005, instruction at the LTC has been primarily directed by international staff through a UNDP project (UNDP 2013a: 10).[21] The 2006 crisis placed a renewed emphasis on establishing order by building a more professional judiciary. By 2012 the judicial system was staffed almost exclusively with Timorese judges. The LTC has produced a significant number of prosecutors, public defenders and, more recently, private lawyers. Courts operated consistently in Dili, Baucau, Suai, and Oeccusse, and fully qualified private lawyers practiced in Dili.

Not surprisingly, challenges remained. The Supreme Court had yet to be established (and still has not been as of 2021). Courts were still backlogged. Case resolution remained time-consuming. Proceedings were often confusing to participants, especially when conducted in Portuguese. While the constitution provided a right to an attorney, legal representation was hard to come by, particularly for those located outside urban centers or facing economic hardship. Public defenders have been known to ask for payment and

[20] The Court of Appeal had been on hiatus since November 2001.

[21] The LTC's curriculum was offered in both Portuguese and Tetum but focused overwhelmingly on Portuguese. Applicants are expected to have an undergraduate degree. While not an official policy, entry structurally favored UNTL graduates. UNTL is considered Timor-Leste's premier university, and courses there are generally taught in Portuguese, unlike those in the other universities, which tended to teach more in Bahasa Indonesian. Even by the end of 2012, admitted applicants often possessed insufficient Portuguese language skills and were assigned to a year of Portuguese language instruction before starting substantive legal study.

were frequently accused of being excessively interested in profit. The number of educational establishments offering degrees in law spiked, but outside of the National University of Timor-Leste (UNTL) their quality tended to be poor. Legal education remained very loosely regulated, and most institutions are for-profit. Real progress had been achieved, but serious challenges remained.

Progress and Problems in Moving Toward the Rule of Law

Timor-Leste still failed to uphold even a "thin" conceptualization of the rule of law. The state legal system's reach is limited. Impunity remains a serious issue (Kent 2012a). Court proceedings are often opaque and take place in Portuguese, a language that participants find incomprehensible without translation. Procedural irregularities are rife. Most people do not know the laws that govern them and cannot easily access the law should they seek to find out, because laws are rarely published in Tetum, the most common local language. As Portuguese laws were often imported wholesale, they often fail to reflect local circumstances. Legal education has improved, but students still learned primarily Portuguese rather than Timorese law at the undergraduate level. And the LTC, at least as of September 2012, remained focused on Portuguese over Tetum. Elected legislators often voted on legislation in Portuguese that the majority of them cannot understand. This raises serious concerns regarding the legislative process and citizens' ability to understand the rules that govern them.

Yet, progress was very real. Timor-Leste's impressive achievements should not be discounted simply because challenges remain. As Timor-Leste's leaders generally embraced rule-of-law ideals, it is worth examining two events that were seen as the biggest potential challenge of those ideals: President Ramos-Horta's extensive use of the pardon power and Prime Minister Gusmão's unilateral decision to release Martenus Bere.

Presidential Pardons

President Ramos-Horta (2007–2012) liberally used the presidential pardon power. He stressed that reconciliation and healing, rather than punishment or retribution, were his top priorities. Most notably, on May 20, 2008, Ramos-Horta unilaterally reduced the prison population by half "when 94 of Timor-Leste's 179 prisoners received a pardon or partial commutation of sentence" (UNMIT 2008). In 2010 he even pardoned individuals implicated in an assassination attempt on his life.

Scholars such as Grenfell argued that Ramos-Horta's behavior "diverges from common understandings of the rule of law in the West, which equate the concept with legal predictability, due process, accountability and individual rights" (Grenfell 2009: 222). These concerns are certainly legitimate, but they reflect quite a narrow understanding of the rule of law and the existence of a very plausible, if controversial, policy rationale for the pardons. It also ignores the practice of states that are closer to meeting rule-of-law ideals. For example, U.S. presidents enjoy plenary pardon power and have even attached elaborate conditions to clemency (Krent 2001). Even before President Trump's provocative use of it (Ford 2020), presidential pardons generated controversy. Crouch, for instance, contends that both Presidents Bush used the clemency power "either to help their own executive branch officials and possibly themselves to avoid judicial prosecution," and President Clinton "sought to excuse financial contributors and aides that, as the president himself, had been subjects of aggressive independent counsel investigations" (Crouch 2008: 722). Yet, few would argue that the United States is fundamentally lawless, since pardon power is exercised in a manner consistent with the law as commonly understood.

As discussed in Chapter 2, the rule of law is best conceptualized on a spectrum ranging from "thin" procedural-based conceptions to "thick" understandings, rather than a binary on/off characteristic (Fukuyama 2004: 59). Even states generally viewed as bound by the rule of law engage in activities that contravene rule-of-law ideals. At the same time, chaotic, corrupt, or limited states, such as the Democratic Republic of Congo, still retain a significant degree of state-imposed legal order (Trefon 2009). The key issue is often intent, and whether a normative commitment exists to the rule of law (Stromseth, Wippman, and Brooks 2006: 76; Fukuyama 2011: 259–260). President Ramos-Horta consistently argued his decisions aimed to promote reconciliation and foster stability rather than interfere with the judicial process (Centre for International Governance Innovation 2011; Robinson 2011: 1015). Even the Justice Sector Monitoring Programme (JSMP), a prominent local NGO that raised serious concerns about the pardons, concedes that Ramos-Horta was primarily interested in fostering reconciliation rather than seeking to undermine the law (JSMP 2009: 14) and that he acted within his constitutional authority (JSMP 2010: 5).

In the constitution, the pardon power is listed under the president's "exclusively incumbent" competencies "to grant pardons and commute sentences after consultation with the Government" (RDTL 2002: Sec. 85(i)). Although the president is required to receive the advice of the government, he is not obligated to heed it. Ramos-Horta's actions were not an abuse of power or a fundamental denunciation of the rule of law, but they still raised serious issues of constitutional design and potential abuse of authority. Extensive use of pardons could sideline the judiciary as the primary institution that structures and guarantees legal order. JSMP, for instance, argued that "wholesale commutation and reduction of sentences could jeopardise efforts to ensure fair and consistent application of the laws in Timor-Leste" (JSMP 2009: 13). One may disagree with Ramos-Horta's justifications, but he was within his legal rights. His successor, President Taur Matan Rauk (2012–2017), was far more restrained. Even more promising, legislation was passed in 2016 that clarified ambiguities and established standard procedures for exercising the pardon power (JSMP 2017: 51–52). Although abuse of the pardon still poses a threat to the rule of law, serious efforts have been taken to mitigate that danger, reflecting a normative commitment to the rule of law by both the president and the National Parliament.

The Release of Martenus Bere

Prime Minister Gusmão's unilateral release of Martenus Bere is arguably the most serious indictment of a Timorese state leader's commitment to the rule of law. Bere, a militia leader, was linked to the deaths of thirty unarmed civilians and three priests in September 1999 (Robinson 2003). In 2009 Bere was captured upon his return to Timor-Leste, where he faced prosecution. Indonesia requested that Bere be released and extradited. Indonesia also implied that failure to do so could damage relations between the two countries. Gusmão tried, but failed, to get a judge to release Bere. Gusmão then ordered Bere's release his own authority (Gusmão 2009: 12–13).

The event sparked significant domestic and international concern (Ki-moon 2009: 10; Amnesty International 2011: 6). The opposition Fretilin party lodged a parliamentary no-confidence motion. Gusmão accused Fretilin of hypocrisy and touted his record during the struggle for independence. However, the main thrust of his argument was legal. He contended that the constitution's mandate to "maintain special ties of friendship and co-operation with its neighbouring countries" justified his actions (RDTL 2002: Sec. 8(4); Gusmão 2009: 4). Gusmão further argued that the original detention was illegal under existing Timorese law. The prime minister contended it was not a preventative detention, and that "UNTAET's Regulation no. 2000/15 on the 'creation of chambers with exclusive and special jurisdiction over serious offences'" excluded any other courts from handling these matters (Gusmão 2009: 10). Gusmão pursued his legal argument further:

> The Penal Procedure Code, through Decree-Law no. 13/2005 of 1 December, expressly safeguards the regime created by UNTAET's Regulation no. 2000/15 . . . Section 30.2 of the Constitution of the RDTL states that "no one shall be arrested or detained except under the terms expressly provided for by the applicable laws". A detention or arrest must always be submitted to the appreciation of a competent judge within the legal timeframe, which did not happen. Therefore I must ask the Honourable Members of Parliament for Fretilin: Who is violating the Constitution and the applicable laws in our Country after all?

The merits of this argument are shaky at best. It is nonetheless striking that Gusmão casts the discussion in explicitly legalistic terms.

Notably, the Bere incident is the most significant event to date in which the government faced widespread criticism for unequivocally disregarding the rule of law.[22] The rarity of these incidents and the attention they attract suggests that the rule of law is taken seriously. A rule-of-law culture does not mean that there is no corruption or abuse of power by individual officials, including high-level ones. Rather, it means that flouting the law is not the norm, that functioning accountability mechanisms exists, and that flagrant violations spark public condemnation. Notably, nothing else similar to Bere's release has occurred.

The behavior of the larger political system is encouraging. The opposition Fretilin party used the parliamentary mechanism of a no-confidence vote to seek Gusmão's censure. A free and open debate ensued. Gusmão survived the vote, but all indications were that Gusmão would have heeded the result if he had lost. The political crisis was resolved through institutional mechanisms. The courts were consistent in condemning the action. Gusmão faced domestic and international condemnation, suggesting, at a minimum,

[22] The performance of Gusmão's coalition governments was not without controversy. As oil revenues increased dramatically, the government faced allegations that corruption had increased substantially (Lundahl and Sjöholm 2008).

that similar actions in the future would not be without cost. Most important, Gusmão either believed his actions were consistent with the rule of law or at the very least felt compelled to justify them as consistent with the rule of law. This is crucial, because the rule of law requires a normative commitment to those ideals. More generally, key leaders from Fretilin and Gusmão's coalition governments have supported rule-of-law principles. It is difficult to overstate the importance of the fact that elite political actors are increasingly willing to play by the rules of liberal democracy and face real threat of punishment if they break the law.[23]

Inherent Tensions in the Quest for the Rule of Law

While these incidents have been used to question the commitment of Gusmão and Ramos-Horta to the rule of law, they more accurately reflect these leaders' commitment to remaining on good terms with Indonesia, their large and immensely more powerful neighbor, and the need for domestic stability. Such incidents highlight a central tension between the immediate demands of state-building and the rule of law:

> Rulers can enhance their authority by acting within and on behalf of the law. On the other hand, the law can prevent them from doing things they would like to do, not just in their private interests but in the interests of the community as a whole. (Fukuyama 2011: 246)

These issues are not limited to post-conflict or developing countries. Even in well-established states, such as the United States and the United Kingdom, commonly seen as being governed in accordance with the rule of law, the temptation exists to disregard legal requirements, not only for personal gain but in the service of compelling state interests, particularly in matters of national security (Bingham 2011: 133–159). Given the immense challenges facing the nascent Timorese state, it is not surprising that controversial incidents occurred. Rather, it is unusual that they are so rare and even then are clearly linked to plausible public policy goals. Still, longer-term consolidation of democratic governance bound by the rule of law is not guaranteed. Even under the best of circumstances, establishing the rule of law and the inclusive institutions that underpin it is often a lengthy, highly contingent process.

[23] For instance, former Minister of Justice Laboto was convicted of corruption in 2012.

Conclusion

This chapter examined East Timor's long struggle for independence and subsequent development. It argued that state practice and policy was notable for its consistency with the independence movement's vision of both a democratic state and how the state and society relate to one another, which in turn has been a valuable source of legitimacy and authority. State legitimacy and authority are essential for transcending conflict (Call 2008a: 7). By drawing on the legacy of the struggle for independence, including its powerful myths and symbols, and on a vision of a modern state committed to development, the state harnessed the power, legitimacy, and capacity of non-state authorities to an extent far greater than did either Portuguese or Indonesian authorities.

Echoes of Portuguese rule linger in the state's reliance on non-state actors to maintain order locally. The Timorese state, however, more effectively regulates non-state actors. Although Portugal formed tactical alliances with non-state authorities, they were invariably tenuous, and there was a constant threat of rebellion. In contrast, non-state actors are now simultaneously legitimized by the state through elections and actively contribute to the state-building project itself. Timor-Leste has transformed from competitive legal pluralism under Portuguese rule and outright combative legal pluralism under Indonesian occupation to cooperative legal pluralism in an independent Timor-Leste.

Timor-Leste still faces serious problems that could undermine further progress toward the consolidation of democracy and the rule of law, or even erode past gains. Nevertheless, its accomplishments are significant and offer insights into how to transform competitive legal pluralism into cooperative legal pluralism. In short, Timor-Leste shows how a post-conflict state, given the right conditions, can undertake effective judicial state-building against a backdrop of extensive competitive legal pluralism.

Timor-Leste did not make this transition in a political vacuum. The indigenous state-building program operated alongside and with significant assistance from international actors. As argued in the next chapter, international involvement can hinder or even derail promising state-led efforts. International support, on the other hand, can make a positive contribution to judicial state-building in settings defined by competitive legal pluralism. However, the relationship between the non-state justice sector and judicial state-building actors in situations of competitive legal pluralism has important consequences for the success or failure of judicial state-building endeavors. As Tansey argues, "Too often, domestic conditions that follow in the wake of international interventions are assumed to follow as a result of international interventions," but "even in the most authoritative of international missions, domestic political elites retain significant levels of formal and informal political power and can profoundly shape the direction of national politics" (Tansey 2014: 185). In other words, state officials, rather than international actors, inevitably have greatest sway.

5

International Assistance for Cooperation and Exclusion in Timor-Leste

Introduction

In Chapter 4 I argued that Timor-Leste's success in judicial state-building stemmed primarily from the decisions of domestic political elites. These actions, however, did not occur in a vacuum. Even after independence in 2002, Timor-Leste received extensive international support. While international assistance can be helpful, it can also be wasteful, ineffective, or even counterproductive to establishing a democratic state bound by the rule of law. In Timor-Leste, however, donor assistance from the United Nations, the United States, and Australian-funded international non-governmental organizations (NGOs) promoted the rule of law and a transition from competitive to cooperative legal pluralism. In contrast, Portuguese bilateral assistance produced mixed results. It helped establish foundational legal institutions but did so in a way that risked state justice's becoming structurally inaccessible to the vast majority of the population.

Why did different types of international assistance have such divergent results? This chapter argues that international assistance from the United Nations, the United States, and Australia followed a coherent strategy based on sustained engagement that adapted to changing circumstances that comported with state goals. The first priority was establishing a viable state justice sector. Once sufficient progress had been achieved, international assistance began to support constructive engagement with the non-state justice sector while continuing to strengthen the nascent state sector. In contrast, Portuguese bilateral assistance, while enjoying exceptionally strong state support, produced mixed results. Portuguese aid was important for establishing a modern legal system, but the system prioritized ensuring the predominance of Portuguese language over accessibility and failed to engage meaningfully with non-state justice.

Contending Orders. Geoffrey Swenson, Oxford University Press.
DOI: 10.1093/oso/9780197530429.003.0005

Summary of Arguments

While state actors were the most powerful force in judicial state-building and played a primary mediating role with non-state authorities, the international community was an important contributor in both areas. In Timor-Leste the state made progress toward a modern democratic state bound by the rule of law through competitive national elections and the creation of judicial institutions that embraced rule-of-law ideals. Equally important, the state's compelling vision of a democratic state committed to the rule of law, combined with competitive local suco elections, established a cooperative relationship with the non-state justice sector. The international community reinforced positive trends in Timor-Lese: its support subsidized the development of the state justice sector and, particularly after the 2006 crisis, worked to bridge the state and non-state justice systems. These activities largely supported domestic initiatives. While its Lusophone legal system presented real challenges, Timor-Leste constitutes a success. This chapter shows how the successful deployment of international judicial state-building assistance with a sound strategic basis can help advance the rule of law and promote a more cooperative relationship between state and non-state justice actors.

Evaluating the Two Main Strands of International Judicial State-Building Assistance

Assessing judicial state-building efforts is difficult. It is an immensely complex task (Fukuyama 2004: 59), and establishing the rule of law takes decades even under favorable circumstances (North, Wallis et al. 2009: 27). Furthermore, some efforts to promote the rule of law have positive effects, while others do not (Stromseth, Wippman, and Brooks 2006). Thus, as discussed in Chapter 1, my criteria examine whether international assistance has *enhanced the prospects for developing and consolidating the rule of law*. This standard allows a nuanced investigation into the consequences of international involvement without conflating engagement with causation. Taking a process-tracing approach, this chapter examines the two distinct strands of assistance. First, the conventional strand of judicial state-building assistance consists of aid provided by the United Nations and its agencies, the United States, and Australia.[1] Second, there was a

[1] The term "conventional" refers not to the aid's content but rather on the fact that it was not given based on historical, cultural, or linguistic affinity. Portuguese assistance reflects the tight bonds at

Lusophone strand of assistance provided by Portugal that was ideologically committed to promoting Portuguese language and culture.[2] While Portugal is not usually a major player in assistance provided for international development, it is in Timor-Leste. In fact, until 2008 it was the largest single donor country (Portuguese Institute for Development Support 2010: 122). In general, Portuguese aid "strongly focused on six partner countries with which it has historical connections, a shared language and close relationships" (OECD Development Assistance Committee 2011: 11).[3]

There are two major differences between the strands of assistance. Both programming streams sought to work with the state to develop a modern, legitimate justice sector in Timor-Leste. However, Portuguese judicial assistance, and indeed all of its programming, were explicitly predicated on "reconstruction of the Education sector and the consolidation of Portuguese language" (Portuguese Institute for Development Support 2010: 121).[4] Second, Portuguese assistance remained exclusively statist, while conventional interventional assistance increasingly recognized the need to engage with justice forums beyond the state.

The Timorese Constitution enshrines both Tetum and Portuguese as official languages (RDTL 2002: Section 13). However, in practice they are far from equal:

> Portuguese, the de facto legal language of Timor-Leste, is spoken by only an estimated seven percent of the national public. Although Tetum is spoken by over 80 percent of the national public, many in government and elsewhere see it as a "trading" language lacking the terminology for legal settings. The result is that laws, judgments and even court orders are

the "government and personal levels, and in trade, migration flows and the strong presence of the Portuguese private sector" (OECD Development Assistance Committee 2011: 25).

[2] The realms of Portuguese and non-Portuguese assistance are not hermetically sealed. For instance, Portugal has been a contributor to the U.N. missions and UNDP activities and has actively pushed for UNDP personnel working in the legal sector to be Portuguese or at least Lusophone. International NGOs have worked to ensure that their activities are not duplicated or opposed by Portuguese-funded actors or programs.

[3] The other major aid recipients are Angola, Cape Verde, Guinea Bissau, Mozambique, and São Tomé–Principe.

[4] The emphasis on promoting the Portuguese language has been overt and formalized. The first two bilateral aid agreements between Timor-Leste and Portugal "place[d] as the top priority the reintroduction of Portuguese into East Timor's education system from kindergarten through university" as comprehensively as possible (Fernandes 2010: 127).

> often handed down in Portuguese, especially by international judges. Many court actors, including national judges, private lawyers and members of Parliament are thus unable to understand, act on, or further disseminate legal information. (Everett 2009: 22)

Language politics permeated the entire state-building endeavor. Rule-of-law assistance from the United Nations, the United States, and Australia was largely indifferent regarding the use of Portuguese and accepted the utility of Tetum, which most of the population speaks. In contrast, Portuguese programs favored Portuguese citizens and those from Lusophone countries. Portuguese aid had little interest in Tetum and maintained that the only solution to the challenges raised by a reliance on Portuguese was more language instruction.[5]

This chapter also corrects common misunderstandings regarding international judicial state-building in Timor-Leste. First, it challenges the enduring myth that international rule-of-law assistance was harmful. Chapter 4 demonstrated the Timorese political and legal elites successfully responded to significant challenges while retaining a commitment to democracy and the rule of law. Yet, negative portraits of international assistance to Timor-Leste abound (Bowles and Chopra 2008; Richmond and Franks 2008; Butler 2012). Even Tansey's compelling rebuttal of the idea that the United Nations bears responsibility for Fretilin's initial electoral victory or the 2006 crisis stresses its aim is "not to exonerate the UN or present an apology or defense of international state-building" (Tansey 2014: 185). My argument goes further. This chapter argues that conventional judicial state-building assistance, while certainly imperfect, offered crucial subsidization and aided the domestic-led process of developing a cooperative relationship between state and non-state justice systems.

Second, this chapter corrects the tendency to view international assistance as either one overarching bundle or simply a set of disparate initiatives. The chapter highlights the existence of two major distinct standards of assistance,

[5] This divide was evident in my work with the Asia Foundation's Access to Justice Program, particularly with an initiative that developed textbooks on the law of Timor-Leste in both official languages as well the working language of English. International UNDP staff at the Legal Training Centre were invariably from Lusophone countries. While occasionally skeptical about Tetum as a legal language, since it lacks an extensive technical vocabulary, they were amenable to legal Tetum and to the use of texts in both official languages. In contrast, the Portuguese-funded law faculty at the National University of Timor-Leste was open to incorporating textbooks in Portuguese but had no interest in using the Tetum texts.

conventional and Lusophone. Conventional assistance from the United Nations, and to a lesser extent the United States and Australia, received substantial attention. The large-scale assistance offered by Portugal, and strongly backed by Timorese political elites, is explicitly linked to the Portuguese language. This aid is at least as important as that offered by the United Nations or Australia, but it is rarely analyzed. Notably, none of the authors cited above discusses the role of Portuguese assistance in any detail. Yet, as this chapter will demonstrate, understanding Portuguese assistance is crucial to evaluating international aid to Timor-Leste.

This chapter addresses conventional and Portuguese assistance in turn. First, it argues that despite inefficiencies and some underperforming projects, conventional international assistance made progress toward building the rule of law against a backdrop of low human-resource capacity, minimal infrastructure, and extraordinary legal pluralism. Second, since independence, international support has helped bolster and institutionalize state-led gains by funding and supporting government priorities. International donors' overarching approach was both strategically sound and adaptive to changing circumstances. From 2002 to 2012 external assistance broadly comported with the state's vision of a judicial system and a state committed to the rule of law. Through process tracing, this chapter examines the major international programs and initiatives to illuminate how international support helped facilitate a transition to cooperative legal pluralism.

While Portuguese aid enjoyed strong domestic political support from both major political parties, the results have been uneven. By 2012 a functioning Lusophone justice system existed. Although more than ten years of efforts to establish Portuguese as a commonly spoken language achieved little (Taylor-Leech 2013), specific Portuguese legal initiatives achieved many of their desired outputs. At the project level, the successes are clear. However, it is unlikely that the nascent Lusophone legal institutions will ever be able to fully support the rule of law as long as they remain largely disconnected from society at large.

Chapter Overview

This chapter is organized thematically rather than in strict chronology. The first section looks at major conventional rule-of-law initiatives and whether each program enhanced the prospects for developing and consolidating the rule of law on a micro-level.[6] The second section analyzes the strategy and trajectory of conventional international judicial state-building activities holistically. The final section discusses Portuguese bilateral support to the justice sector and its consequences.

Conventional International Assistance (2002–2012)

After the country's independence in 2002, international involvement continued far beyond Timor-Leste's economic or strategic importance. Throughout this period the United Nations, the UNDP, and the USAID-funded programs, most notably the Asia Foundation's Access to Justice (ATJ) program, and Management Sciences for Development's (MSD) Justice Institution's Strengthening Program (JISP). Australia established the Justice Facility Program after Timor-Leste emerged as a major regional security concern after the 2006 crisis.[7] While some initiatives proved underwhelming, such as efforts to pursue transitional justice and draft legislation stipulating the content of non-state justice, overall international assistance helped establish a functioning nationwide justice system that reflects rule-of-law ideals and constructively engages with non-state justice.

The United Nations

From 2002 to 2012 there were three distinct U.N. missions in Timor-Leste. The UN Mission in Support of East Timor (UNMISET), from May 2002 to May 2005, sought to support Timor-Leste's transition to democracy and the rule of law (UNSC 2002b: 2). After independence, the Timorese state was initially eager to "share[] its sovereign prerogatives with UNMISET" to improve state administration, especially in the judiciary (Zaum 2007: 193).[8] This

[6] As discussed in Chapter 3, the justice sector should be understood as encompassing both state and non-state judicial activities. Certain actors, most notably suco chiefs and the police, have one foot planted firmly in each camp. Nevertheless, for the sake of conceptual clarity, it is worth examining initiatives that are primarily targeted at the state and activities aimed at non-state actors separately, as determined by programmatic intent.

[7] The World Bank and other organizations also played smaller roles. For more information about other international rule of law assistance, see Freedom House and American Bar Association Rule of Law Initiative (2007), U.N. Independent Comprehensive Needs Assessment Team (2009), and RDTL (2010).

[8] Timor-Leste assumed formal control over its own security on May 20, 2004.

included placing U.N.-funded international advisors directly in Timorese judicial and legal structures, including international judges and advisors in the Ministry of Justice, as well as responsibility for security (UNSC 2002a: 4).

The UN Office in Timor-Leste (UNOTIL) operated from May 2005 to May 2006. UNOTIL reduced overall expenditure but still prioritized support of the rule of law (UNSC 2005: 1). U.N.-funded international experts remained embedded throughout the judicial sector. In 2005 the presence of outside experts prevented the justice system's collapse when all local justice sector personnel failed their examinations and were subsequently suspended. As a result, the justice system was temporarily run solely with international personnel until local personnel could be sufficiently qualified (Jensen 2008: 133).

After the 2006 crisis and the return of international peacekeepers, the United Nations once again invested heavily in Timor-Leste. In August 2006 the Security Council authorized the UN Integrated Mission in Timor-Leste (UNMIT) (UNSC 2006). UNMIT's mandate would be extended until December 31, 2012.[9] As building a viable, legitimate state justice sector again became a top priority, international presence in the justice sector expanded, and an administration of justice support unit was created in 2007. The nine-person unit sought to help build national capacity and improve justice sector coordination by working with state officials, embassy staff, and international NGOs (UNSC 2011).

Transitional Justice

After independence the United Nations, and the broader international community in Timor-Leste, supported transitional justice to punish perpetrators of violence.[10] The United Nations officially backed justice for victims of the occupation and violence related to the 1999 referendum (Annan 1999a). Transitional justice initiatives, however, produced decidedly mixed results and exposed the limits of international influence.

UNTAET established a Special Crimes Investigation Unit (SCIU) and Special Panels to address these issues (UNTAET 2000a, b). Funded and directed by the United Nations, the SCIU had four teams "each staffed almost exclusively by international prosecutors, investigators, and case managers" (Katzenstein 2003: 251). SCIUs operated until 2005 but produced

[9] The U.N. police also assumed direct responsibility for security until 2011.

[10] Transitional justice is often viewed as vital to promoting the rule of law and long-term stability, but its empirical consequences are subject to debate (Mendeloff 2004; Thoms et al. 2010; Reiter, Olsen et al. 2013; Swenson and Kniess 2021).

unimpressive results. During its tenure, 101 people came before the panels, and eighty-seven received final decisions (Reiger 2006: 151). The eighty-one defendants convicted were almost exclusively low-level domestic actors (ibid.). The SCIU did little to bring those actually responsible for the planning and execution of mass killings to justice (Kent 2012b). Its lackluster performance must be understood in light of state officials' deep skepticism. While operations continued after independence, state authorities did not actively support SCIU's work and often hindered it (Cohen 2007).

In contrast, the government fully backed the UNTAET-initiated Commission for Reception, Truth and Reconciliation (CAVR). The CAVR sought to facilitate reconciliation by compiling a definitive record of events, providing survivors an opportunity to be heard, and fostering social reintegration (UNTAET 2001). The CAVR process was well regarded, particularly for incorporating localized non-state justice approaches (Nixon 2012). CAVR's mandate, however, was decidedly non-retributive, and policy makers flatly rejected its calls for prosecution (Kent 2014).

Since independence, state officials have prioritized good relations with Indonesia over transitional justice (Robinson 2011). Then–Foreign Minister Ramos-Horta explained the government's stance in 2004: "The government of East Timor does not contemplate lobbying for an international tribunal to try the crimes of 1999 because we know this would undermine the existing relations between the two countries" (Ramos-Horta, quoted in Global Policy Forum 2004). Ramos-Horta was not alone. President Gusmão and the larger political establishment were equally uninterested in the prosecutions related to violence during the occupation and independence referendum (Kent 2014). After the 2006 crisis, the United Nations established the Serious Crimes Investigation Team. This unit, however, met the same fate as its predecessor and failed to deliver any notable results.

No Indonesian officials were ever successfully prosecuted for the violence surrounding the independence referendum. The prospect for future trials is bleak. Calls from the United Nations and the international community for transitional justice waned. The meager results of transitional-justice efforts highlight the United Nations' limited ability to pursue change when opposed by powerful state officials. The fundamental differences were stark. The ideological underpinnings of the United Nations' initiatives were decidedly retributive. This approach enjoyed significant popular support. However, as discussed in the last chapter, Timorese political elites strongly favored a reconciliation-based approach to both the violence surrounding independence and the 2006 crisis. While controversial, this belief has an internal logic. This reconciliation model has been successfully utilized in certain cases, for instance in Spain after Franco, though it is by no means guaranteed to succeed (Posner and Vermeule 2004). The United Nations was able to establish, fund, and even direct entities to pursue retributive transitional justice. There were prosecutions and even some convictions. The U.N.-backed CAVR process was widely considered a success because it generated a definitive narrative of the conflict and allowed people a meaningful right to be heard.[11] Of course, the CAVR was fundamentally reconciliatory in nature. As such, it enjoyed strong state support. In a context where retributive transitional justice was viewed as a national security threat at the highest levels of government, the United Nations' ability to pursue transitional justice was heavily circumscribed.

The UNDP

The UNDP justice sector aid started in 1999 and the agency's Strengthening the Justice System Program (JSP) commenced in 2003. At first UNDP prioritized underwriting advisors in key legal institutions, such as prosecutors and public defenders' offices, the courts, and the MOJ, along with drafting educational material and providing technical assistance (Frigaard, Mullaly et al. 2008; UNDP 2013b). UNDP's work later broadened significantly, both in response to the 2007 midterm evaluation and external events, most notably the 2006 crisis and the new government after the 2007 parliamentary elections.[12] Assistance to the Legal Training Center (LTC) became a focal point after 2006: the UNDP funded six international lecturers and numerous Portuguese-language instructors (UNDP 2011a: 109).[13] While the LTC officially operated under the auspices of the MOJ, the director generally deferred to the entirely Lusophone UNDP staff (International NGO Manager

[11] While an opportunity to share their stories and hear the truth is often viewed as beneficial in and of itself, it remains unclear if this process actually fosters relief, healing, or forgiveness among participants or helps the peace-building process (Mendeloff 2004).

[12] While noting some advances, the evaluation found that "overall progress on the achievement of the programme outcomes has been slow, and, as yet, the programme has not made a significant impact on access to justice in Timor-Leste" (Frigaard, Mullaly et al. 2008: 5).

[13] Students who were accepted, particularly those who did not graduate from the Lusophone National University of Timor-Leste (UNTL), were often required to undertake intensive Portuguese-language training before commencing their studies.

2014). Moreover, UNDP instructors enjoyed wide leeway regarding administration, curriculum, and instruction. The LTC quickly became the educational epicenter of the legal profession. Going forward, all indigenous judges, prosecutors, public defenders, and private lawyers were required to be LTC graduates. Although an eventual transition to Timorese operational control was envisioned, the LTC depended on foreign assistance through 2012.

The UNDP also attempted to determine and codify the relationship between the state and non-state justice sectors. In other words, it is a paradigmatic example of the harmonization and incorporation approaches discussed in Chapter 3. Following the 2006 crisis, the new minister of justice expressed interest in a "UN proposal on traditional justice mechanisms and their relationship with the formal justice system in Timor-Leste with a view to developing a legal framework" (UNMIT, quoted in Nixon 2012: 197). Consultations were held. At least two versions of a draft law were completed, but they not circulated outside the U.N. system and MOJ's upper echelons. The MOJ, however, did not endorse either version, nor did officials offer clear guidance on what a suitable law should entail. The UNDP subsequently abandoned attempts to regulate non-state justice (UNDP Official 2014). The domestically led Community Authorities Law, which mandated the election of suco chiefs and delineated the scope of their authority, thus remained the most important regulation of the non-state justice sector.

Australian- and U.S.-Funded NGOS

After Timor-Leste achieved independence, international NGOs provided support, but activities ramped up after the 2006 crisis "and the subsequent rethink" of rule of law assistance (Government of Australia 2007: i). The biggest funders were the United States and Australia. Through USAID, the United States supported judicial state-building in Timor-Leste post-independence. Moreover, U.S. assistance to Timor-Leste, unlike that to Afghanistan, benefited from a lack of competing strategic interests that risked compromising its commitment to promoting democracy and the rule of law (Girod 2015). For its part, Australia did not invest heavily until 2008, but afterward it overtook Portugal as the single largest funder. This section examines three major implementing NGOs: MSD, the Asia Foundation, and the Justice Facility, in turn.

Justice Institution's Strengthening Program (2005–2012)
In May 2005 USAID awarded MSD the contract for the JISP. Program activities continued until 2012 (Management Sciences for Development 2012). Based on the premise that enhancing institutional capacity was essential for a viable state justice sector, the JISP sought to "build[] administrative and management capacity in Timorese justice sector institutions" with a particular focus on "training and technical support in general administration, financial management, human-resource administration, good governance and anti-corruption practices" (Management Sciences for Development 2010: 2). The JISP supported the law courts, the MOJ, prosecutors, public defenders, the Provedor for Human Rights and Justice (PDHJ), and the Anti-Corruption Commission (ACC) (see Management Sciences for Development 2012 for a complete overview). MSD and its evaluators were particularly proud of its financial management training for major justice sector institutions (Chopra, Pologruto et al. 2009: 19). The JISP also adapted to institutional changes in how corruption was addressed, most notably the shift in emphasis from the PDHJ to the ACC. The extent of corruption is difficult to determine, but most people believed that corruption had increased dramatically and that a more robust institutional response was required (Soares 2013). The program worked with the ACC and other agencies on their core financial, administration, auditing, and other technical competencies. These technical capacities are necessary though by no means sufficient for state organizations to combat corruption (Rose-Ackerman 1999).

The Asia Foundation—Access to Justice Program (2002–2012)
The most significant, long-running justice sector program was the Access to Justice (ATJ) program, implemented by the Asia Foundation and funded by USAID. ATJ provided free legal aid in partnership with local legal aid NGOs seeking to resolve disputes "through the formal justice sector or through alternative dispute resolution that respects human rights" (Asia Foundation 2003: 3).[14] Building legal aid providers' institutional capacity was another major priority. ATJ "trained the legal aid organizations in substantive knowledge, legal skills and institutional strengthening, which include administrative management and financial management" (Coghlan and Hayati 2012: 15–16). Moreover, the program addressed gaps in state-sponsored legal

[14] The Asia Foundation also engaged with state and non-state authorities in different capacities through its Local Governance, Elections, and Civil Society Programme as well as community policing programs.

services by offering mobile legal support in rural areas as well as defending clients in state court proceedings that public defenders were unwilling or unable to represent. Over 6500 people received assistance through the ATJ program's legal aid activities between 2002 and 2012 (Asia Foundation 2012: 28).

Non-State Justice

In the wake of the 2006 crisis, in 2008 ATJ began supporting paralegal programs (Asia Foundation 2009: 24–26). Paralegals are "lay people with basic training in law and formal government who assist poor and otherwise disempowered communities" (Maru 2006: 429). Through its network of paralegals in Baucau, Manatuto, Viqueque, Lospalos, and Oecusse, the ATJ program sought to interact directly with non-state judicial authorities as well as bolster the capacity of paralegals and suco council members. Paralegals served a number of roles. They referred individuals who either were unaware that the state justice existed or were unsure how to access it; they resolved minor disputes themselves if the participants preferred; or they assisted with the suco dispute-resolution process (Coghlan and Hayati 2012: 32–33). This bridging role was vital. For instance, all eighteen suco chiefs in the remote district of Oecusse stated that ATJ-funded paralegals were their only link to the state justice system and the police (Graydon 2011: 33). During the four years of active programming, paralegals assisted with 3110 cases (Asia Foundation 2012: 29). Simultaneously, there was general recognition that non-state actors had little knowledge of state laws. After a wholesale evaluation in 2010, the Asia Foundation undertook a comprehensive paralegal training program on the most important aspects of applicable state law. Training also sought to ensure that suco dispute resolution was generally consistent with broad due-process and human-rights norms.

Education and Legal Civil Society

Knowledge of the law by both legal professionals and society at large is essential for developing the rule of law. Thus, education was another focal point. The Asia Foundation partnered with UNTL to develop the first textbooks on the law of Timor-Leste. Textbooks were produced in both official languages. Previously, all Timorese law schools relied on textbooks on Portuguese or Indonesian law. ATJ supported the translation of laws into Tetum, produced guides on key legal issues in Tetum, and organized public consultations and socialization of recently enacted laws and draft legislation.

ATJ funded legal awareness and advocacy NGOs, most notably the Judicial Systems Monitoring Program (JSMP), from 2002. JSMP closely observed the justice and legislative sectors. It produced both qualitative and quantitative assessments of the justice system, including reports, policy briefs, and Tetum translations of draft legislation. Because the government rarely published official statistics, these were vital reference tools. As a long-standing, respected local NGO, JSMP positively influenced legal and political discourse.[15] JSMP stressed the need to address impunity concerns and gender inequality issues. For instance, JSMP highlighted President Ramos-Horta's extensive use of the presidential pardon power and proposed ways to address the issue through legislation. Notably, in 2016 these efforts culminated in the passage of a law on presidential pardons (JSMP 2017).

Justice Facility (2008–2012)

The primary vehicle for Australian rule-of-law aid was the Justice Sector Support Facility Programme (Justice Facility) implemented by GRM International. It initially focused on the overarching goals of "Corporate Management Support for Core Institutions" and "Civil Society demand for Justice" (Justice Facility 2011: 6). In 2010 the focus was shifted slightly to "Institutional Development" and "Civil Society Support and Access to Justice" to "more closely align with GoTL [government of Timor-Leste] priorities" (Justice Facility 2011: 6). Justice Facility prioritized technical assistance. They funded external experts within legal institutions, a new comprehensive justice sector information management system, and the development of the Justice Sector Strategic Plan 2011–2030. They also supported numerous legal actors through a small grant program as well as institutional capacity-building support.

When it comes to establishing the rule of law in developing countries, "the legal profession plays a role in safeguarding liberties against incursions by the state," as well as in providing education to practicing lawyers and the public at large (Daniels and Trebilcock 2004: 116). Because an inclusive, competent professional association is important to consolidating the rule of law, Justice Facility supported the Assosiasaun Advogados Timor-Leste (AATL). Upon

[15] For example, La'o Hamutuk and Belun were recognized for high-quality research and advocacy but faced criticism over being perceived as foreign.

reaching a sufficient number of fully qualified attorneys, AATL was slated to transform into the official bar association of Timor-Leste, at which point it would also assume significant gatekeeping and regulatory powers over the legal profession.

Programmatic Progress and Challenges

Overall, these programs and measures had a positive effect and bolstered the domestic judicial state-building process. Nevertheless, they faced significant challenges. The impact of U.N. efforts to support transitional justice has been underwhelming. JSP's midterm evaluation noted that while it made a "significant contribution to the strengthening of the justice system," progress has been slower than anticipated (Frigaard, Mullaly et al. 2008: 5). Moreover, UNDP's non-state justice legislation was criticized by both scholars and practitioners for trying to impose uniformity across a highly diverse system. To UNDP's credit, legislative drafting efforts were jettisoned once it became clear that these plans were not feasible and lacked sufficient state support.

While praising the small grant program and the integrated justice sector information system, the independent 2012 review of Justice Facility featured serious criticisms. Evaluators determined that the program was overly ambitious and therefore often failed to complete planned projects (Peake, Pearce et al. 2012: 16–20). Justice Facility was further faulted for poor internal monitoring, substantial management shortcomings, insufficient strategic planning, and programmatic sustainability (ibid.).[16] Reviews of MSD and the Asia Foundation programs were quite positive (Chopra, Pologruto et al. 2009; Graydon 2011; Coghlan and Hayati 2012). Nevertheless, as could be expected given the fluid programming environment and the inherent challenges of international development work, these programs faced significant obstacles. The Asia Foundation suspended collaboration with numerous local legal aid partners after its internal monitoring activities discovered financial irregularities (Asia Foundation 2012). For its part, MSD was forced

[16] Not surprisingly, the program's implementers did not accept the evaluators' criticisms unequivocally. While open to some suggested changes, the Justice Facility challenged the claims regarding the program's management, efficiency, and monitoring and evaluation (AusAID 2012).

to abruptly end their programming in the vice prime minister's office after his resignation (Management Sciences for Development 2012: 10).

Conventional international assistance certainly left room for improvement. Still, it addressed compelling needs in a way that bolstered rather than undercut domestic efforts. Initiatives were on the whole strategically sound and supported the state's goals, which increased their impact. Supporting the state's goals does not necessarily generate positive results, but in Timor-Leste state and international actors shared a common goal of establishing an effective justice system. As the next section highlights, international assistance helped establish a viable state justice sector with nationwide reach.

Conceptualizing Conventional International Judicial State-Building in Timor-Leste

International support to Timor-Leste evolved between 2002 and 2012. As discussed in Chapter 4, at independence Timor-Leste was characterized by competitive legal pluralism. The state and non-state justice sector actors were open to limited engagement but remained mutually skeptical, as the central state was conflated with the U.N. administration. After independence, however, the state was able to transform the predominant dynamic from competitive to cooperative legal pluralism, where the state and the non-state justice sector worked together toward a shared vision of a modern Timorese state. This Herculean task was accomplished by translating the legitimacy of the independence movement and the state structures it envisioned into an actual state itself that embraced rule-of-law ideals while simultaneously rooting the local political and legal order in democratic choice.

Initially, aid focused on bolstering state justice sector institutions in a straightforward subsidization strategy. After the 2006 crisis, however, international actors realized that meaningfully engaging the non-state justice sector was also essential. This was no easy task. Localized non-state dispute resolution mechanisms, most notably dispute resolution by suco leaders, often produced results that seemed legitimate to local populations but fell short of international due-process and human-rights norms. Assistance took a flexible yet principled approach that sought to improve the quality of non-state justice and links between the two sectors while helping construct a modern judiciary. Collectively, this support helped reinforce the state-led trend toward cooperative rather than competitive legal pluralism.

Subsidization Strategy from Independence to the 2006 Crisis

From 2002 to May 2006 international assistance supported the creation and maintenance of basic state justice institutions. The underlying rationale was obvious. For most citizens, the state justice sector was more an abstract concept than an everyday reality. It faced compelling human-resource, training, infrastructure, and material needs that international aid tried to address (Marriott 2009; RDTL 2010a). While this approach received substantial criticism, it was consistent with the views of Timorese political leaders across the political spectrum. Assistance was concentrated in Dili and Timor-Leste's three other cities with courts, Baucau, Oecusse, and Suai.[17] Additional assistance tried to help these organizations become more professional and better uphold the norms of due process, human rights, and the rule of law. In other words, international assistance emphasized subsidies to aid state institutions but largely eschewed the non-state justice sector.

Although international assistance faced criticism for focusing excessively on state institutions and supposedly foreign notions of democracy and the rule of law, these objections miss the mark. State institutions were by far the area of greatest need precisely because they were new and needed to establish their legitimacy. Frustration from the international community over the perceived slow pace of institutional progress, while understandable, fails to recognize that building inclusive institutions underpinned by the rule of law is a slow and daunting process. As previously discussed, these institutions were not some foreign imposition. Rather, they were a legitimate expression of the independence movement's aspirations. The independence movement was explicitly predicated on a commitment to democracy and the rule of law. This is the type of assistance that international actors can best deliver (Carothers 1999).[18]

Unless one envisions a state underpinned exclusively by non-state law, a prospect not even the most radical critics of international efforts endorse, a functioning state justice system is necessary in its own right as well as a means to engage non-state authorities. While subsidization efforts faced

[17] A notable exception was the Asia Foundation ATJ program's mobile lawyers initiative, whereby legal-aid lawyers traveled the districts, as well as the paralegal program established in 2008, which focused on local justice (Asia Foundation 2012). Likewise, a smaller Avocats sans Frontières program established local paralegal networks from 2005 to 2007 (Low 2007: 6).

[18] The rationale was not purely magnanimous. Codifying the dominance of Portuguese over English or Indonesian placed Lusophone political elites and those able to successfully assimilate in a dominant position.

some implementation issues, it did enhance prospects for the development and consolidation of the rule of law in Timor-Leste. Other approaches may have been possible. For instance, international assistance could have promoted bridging the state and non-state justice sector from the outset, but it is extremely difficult to create a sustainable relationship between a highly developed system and one that is in an embryonic form.

Even in the early years after independence, non-state justice was recognized as important. Non-state justice, however, was not a programmatic focus. It presented obvious programmatic challenges: non-state justice is inherently diffuse and highly localized. Even geographically close communities can vary significantly in how disputes are resolved and the norms of conduct (Nixon 2012). Although non-state justice enjoyed local legitimacy and functioned more or less smoothly, many international actors believed that these forums violated human-rights norms and discriminated against women (Hohe 2003). Beyond establishing the electoral framework, high-level state officials were equally uninterested in directly engaging the non-state justice sector. They were eventually swayed by the 2006 crisis and the growing recognition that a workable state legal system required buy-in from non-state actors (Legal Adviser from the RDTL President's Office 2014).

By early 2006 Timor-Leste was widely seen as an international judicial state-building success story; consequently, international aid was slated to decrease dramatically. For instance, then–World Bank president Paul Wolfowitz claimed Timor-Leste had "opted for peace" and "come together as a nation where so many other countries fall apart in factions" less than a month before the crisis erupted (Wolfowitz, quoted in Sahin 2007: 250). The reality on the ground, however, was far more tenuous, and clear evidence suggested that the country was less stable than many international observers believed (Goldstone 2012).

Diversified Approaches in Conventional International Assistance post–2006

The 2006 crisis was a seminal event in Timor-Leste's domestic politics and a major influence on international judicial state-building assistance. Timor-Leste again topped the United Nations' agenda alongside a massive influx of new international support. The rule of law emerged as a paramount concern. The U.N.'s Independent Special Commission of Inquiry, for instance, argued the crisis "can be explained largely by the frailty of State institutions and the weakness of the rule of law" (UN Independent Special Commission of Inquiry for Timor-Leste 2006: 2).[19] The crisis served as a stark reminder that building an effective state judicial system bound by the rule of law takes many years and remains contingent in circumstances far more favorable than those that obtained in Timor-Leste (North, Wallis et al. 2009).

Since the crisis, state institutions have made significant progress, including the creation of a viable, if understaffed, cadre of judges, prosecutors, public defenders, and private lawyers. By 2012 there were thirty-one judges, twenty-four prosecutors, and twenty-two public defenders (JSMP 2012). Popular faith in the justice system likewise increased. According to a major nationwide survey, 88 percent of the population expressed confidence in the state system (up from 77 percent in the previous, 2008 survey), while 94 percent expressed confidence in the non-state system (Marx 2013: 15). The geographic reach of state justice expanded. Regional courts in Baucau, Oecusse, and Suai became fully and consistently operational, in contrast to the situation in the immediate post-independence period, when closures were common (JSMP 2004). Enhanced reach and robustness made the state justice sector an increasingly potent force in dispute resolution. The non-state justice sector remained the most commonly used venue, but an ever-increasing number of citizens experienced a meaningful choice between the two systems. Still, significant gaps persisted. There were too few public defenders to meet popular demand (Asia Foundation 2012: 16). International assistance, however, helped through the provision of free legal aid by local NGOs, international public defenders, and training to expand the number of qualified legal personnel.

Expanded Scope for Subsidization

Subsidization remained a key strategy but evolved to include the provision of legal education and legal services. Assistance was aimed at bolstering public understanding of the legal system and state laws. In other words, expanded subsidization efforts included the rule of law's cultural aspects and increased attention to professional and educational institutions such as AATL, the LTC, and UNTL. These kinds of knowledge and beliefs are vital. The RAND Corporation's comprehensive survey of international nation-building efforts

[19] A joint assessment by Freedom House and the American Bar Association echoed this finding (Freedom House and American Bar Association Rule of Law Initiative 2007: 15).

since World War II found that "a widely shared cultural commitment to the idea of rule of law" is essential (Dobbins, Jones et al. 2007: 88).

The 2006 crisis sparked a renewed emphasis on passing or revising legislation related to the state and non-state justice sectors, most notably the Private Lawyers Law, the Penal and Civil Codes, and the 2009 revisions to the Community Authorities Law, which provided a domestic legal framework and ended reliance upon outdated Indonesian law. External funding underpinned legal educational institutions, support for legal professional associations, and technical assistance for legislative drafting. International aid likewise sought to socialize the general population regarding key pieces of legislation, such as the draft laws on legal aid and land (Marriott 2012; Asia Foundation 2012).

Emergence of a Complementary Bridging Strategy

After the crisis, working constructively with non-state justice authorities was seen as essential for long-term stability (International NGO Manager 2014a; International NGO Professional 2014b; Legal Adviser from the RDTL President's Office 2014). Because most disputes were still settled outside state courts, this approach demanded significant engagement with the non-state justice sector. A bridging strategy became important, particularly for the Asia Foundation's ATJ program. As the program's final report explains, since 2006 "the program's priorities shifted to serving as a bridge between formal justice sector institutions . . . and informal dispute resolution mechanisms" (Asia Foundation 2012: 11). This approach emphasized improving the quality of non-state justice outputs and capacity-building for its leaders, ensuring access to the state system, and increasing public knowledge about RDTL law and the state legal system (ibid.: 13–14).

A bridging approach funded paralegals affiliated with legal aid organizations. Paralegals were drawn from the communities that they served and performed three main tasks. First, they worked to ensure that all serious criminal matters, including domestic violence, were sent to state courts and lawyers were available to handle civil matters. This was accomplished by linking paralegals with legal aid organizations and state judicial actors. Second, paralegals offered advice and technical assistance to help make non-state justice processes fairer and respectful of basic human-rights norms, most notably rudimentary forms of due process and equity before the law. Finally, paralegals provided referrals and information about state law and the state legal system for local communities (Swenson 2018b). Post-2006 international assistance helped people access state courts and make an informed choice regarding how to resolve their disputes.

International assistance largely avoided harmonization and incorporation strategies. A notable exception was UNDP's two draft customary laws, which sought to codify non-state law. However, international efforts to standardize non-state justice quickly proved infeasible without state support and the existence of divergent customs across the country. Moreover, there was no pressing need for a non-state justice law, as most serious cases were supposed to be referred to state courts under the existing community authorities law.

Synergy with State Priorities

The 2006 crisis influenced the design, outputs, and delivery of international assistance. In turn, key domestic political and judicial actors increasingly recognized the need to engage with non-state justice. Initially, alignment with state priorities simply meant establishing functional institutions. As the state legal system became increasingly sophisticated, programming options proliferated. All major international actors stressed that their programs were collaborative and consistent with official goals (Government of Australia 2007; UNDP 2008; Management Sciences for Development 2012; Asia Foundation 2012).

The increased alignment of international assistance with the Timorese government's objectives was necessitated in part by the discovery of petroleum resources. The early Fretilin government had a modest national budget, leaving it dependent on international assistance. Once oil revenues came online, state funds increased dramatically. In 2005 Timor-Leste's annual budget was US$75 million, but by 2006 it had grown to US$330 million (Bowles and Chopra 2008: 290). In 2010 "95 per cent of the national budget [was] dependent on gas reserves in the Timor Sea" (Lothe and Peake 2010: s434), and these reserves continued to boost the state budget, which, in 2012, stood at US$1.76 billion (RDTL 2011). This dramatic increase enabled the government to start declining projects that did not reflect its priorities, including transitional justice initiatives.

Overall Impact of Conventional International Assistance

Despite its limitations, international judicial state-building assistance enhanced the prospects for developing and consolidating the rule of law. It also promoted a workable relationship between the state and non-state justice sector by both supporting and improving the quality of legal institutions as well as helping link the state and non-state justice sectors based on citizen choice and relevant law. Furthermore, it is to the credit of international actors that they resisted the temptation to directly impose their preferences and instead supported state-driven plans even when they did not fully agree with those plans. International assistance helped promote a rule-of-law culture more broadly by supporting institutions such as the LTC and AATL but also by expressing concerns when high-level state officials' actions raised questions about their commitment to the rule of law, such as Gusmão's unilateral decision to release Bere.

Portuguese Bilateral Judicial State-Building Assistance

Portugal was the single most important external player in post-independence Timor-Leste. Timorese state legal institutions and much of the law itself are overtly modeled on their Portuguese equivalents. Yet Portugal's role in Timor-Leste has received little attention. The explicit goal of Portuguese assistance is a Portuguese-speaking society nationwide (Portuguese Institute for Development Support 2010). Portugal promoted the selection of Portuguese as the official language and the continued prevalence of Portuguese in the legal sector and state institutions more broadly (JSMP 2012: 18). Some evidence even suggests that "the Portuguese government has made its grants conditional on favouring the Portuguese language" (Marriott 2012: 67). Portuguese language preferences had strong high-level government support. Portugal's goals even shaped the United Nations' work as it lobbied for major U.N.-funded positions to require fluency in Portuguese. This favors Portuguese candidates as well as applicants from other Lusophone countries, even if those personnel are otherwise less qualified.[20]

Portuguese assistance likewise focused on subsidization, albeit a particularly extreme version thereof. Portuguese aid is a rare example of a bilateral donor shaping an entire country's legal landscape. The Lusophone legal system, however, was Dili-centric and showed minimal interest in justice beyond the state even after the 2006 crisis. This tendency remained even though the language remained unintelligible to the vast majority of the population

[20] While they operated in relative isolation, Portuguese-supported programs were willing to work with other, non-Lusophone actors, particularly when it was clear the program did not aim to supplant Portuguese as the primary legal language.

and the Portuguese-speaking legal elite was becoming more and more disconnected from society at large. Portuguese aid providers never wavered in their vision of a Lusophone Timor-Leste, but they were unwilling or unable to commit the resources to make that vision a reality.

Goals and Successes of Portuguese Judicial State-Building Support

Portuguese judicial state-building emphasized placements of Portuguese advisors or regular staff in state bodies, including judges, prosecutors, and public defenders. These actors enjoyed immense influence and autonomy as they frequently advised state officials with limited understanding of Portuguese. Portuguese legislative drafters played a primary role in crafting most important legislation, including in the Parliament as well as in the legal drafting unit of the MOJ. From its inception, the law faculty at the University of Timor-Leste (UNTL) operated under extensive Portuguese influence (Nixon 2012: 161). The law curriculum was directly controlled by the Portugal-based Scientific Curriculum Commission. The faculty consisted entirely of Portuguese instructors until the end of 2010, and Portuguese influence was still paramount through 2012.

Though controversial with the general public, these programs enjoyed robust support from high-level state officials. Linguistic ties were very deep. During the Indonesian occupation, "83% of the first Constitutional Government's ministers had been living in exile, primarily in Lusophone countries" (Engel 2007: 13). The constitution and the civil law legal system were explicitly modeled on Portugal. Laws not only drew heavily on their Portuguese counterparts but were often copied wholesale. Portuguese experts were well placed to assist with legal interpretation. This dynamic became self-perpetuating, as the international legal drafters within parliament and the relevant ministries were overwhelmingly Portuguese.

The justice system's transformation has been striking. At the time independence was first declared, the state legal system was completely devastated in physical infrastructure and woefully lacking in human resources. The use of Portuguese was sparse. A major Asia Foundation survey conducted in December 2002 found that Portuguese was spoken by only 7 percent of the population, who "were mostly older, educated, higher income Dili residents," and read by 10 percent (Asia Foundation 2004: 8).

By the end of 2012, much had changed. There were courts operating nationwide, a viable legal profession, and a legal training regime that was decidedly Lusophone in orientation. Laws and regulations were written in Portuguese. Reputable legal education was conducted almost exclusively in Portuguese. Court decisions were in Portuguese. The Dili-based private lawyer community was overwhelming Lusophone. The 2002 constitution formally established Portuguese and Tetum as official languages equal in stature, but even today the Portuguese language and Portuguese institutions occupy a paramount position within the legal system and the state more generally.

The Lusophone legal order is a remarkable accomplishment, given the relatively low levels of Portuguese linguistic penetration into society prior to 1975 and the destruction caused by Indonesia's occupation. However, that legal order risks becoming perilously disconnected from broader Timorese society. After over a decade of independence and sustained state and bilateral efforts to promote Portuguese as a viable, commonly spoken language nationwide, less than 10 percent of the population spoke it (Marx 2013: 34).

Challenges with Portuguese Bilateral Assistance

The Lusophone legal order had an internal logic and elite political support, especially among political leaders who were active during the initial 1970s independence struggle, but it generated real challenges. Conventional international assistance highlighted the importance of state support for external aid to promote the rule of law. Portuguese assistance shows that extensive state support does not necessarily lead to positive results or contribute to establishing and consolidating the rule of law. Timor-Leste is radically different from Portugal. Portugal is a developed country with a high standard of living, excellent infrastructure, the rule of law, and a consolidated democratic state (Fishman 2011). Timor-Leste has a long history of colonialism and occupation, low levels of economic development, a state legal system that is still under construction, and a democracy that remains unconsolidated. The wholesale importation of Portuguese laws and legal norms was ill-suited to the country's needs.

The aggressive pursuit of a Portuguese-style legal system, conducted primarily in Portuguese, isolated much of the population from the state legal order and will continue to do so for the foreseeable future. The problem is nationwide. Despite limited efforts to provide live translation, the use of

Portuguese "continued to be an obstacle in nearly all district courts" (JSMP 2013: 27). Portuguese language and culture form a "distinct and deep divide between the old and young leaders today" (Niner 2007: 114). Society is moving further and further away from its colonial roots and the struggle for independence. As an independent evaluation bluntly stated, "Limited fluency in Portuguese is a barrier to accessing legal education and professional training" and the broader operation of the legal system, which in turn is "hindering the effective functioning of the justice system; undermining due process and, ultimately, limiting access to justice" (Frigaard, Mullaly et al. 2008: 11). Without significant reform, a myopic focus on Portuguese law and language at the expense of the widely spoken lingua franca of Tetum risks establishing a legal system incomprehensible to most of its citizens and excluding many promising individuals from the legal sector.

Language problems are systemic outside the court system. The two primary organs of drafting law, the National Parliament and the MOJ's legislative drafting unit, work in Portuguese (RDTL MOJ Official 2014). In both instances, the laws are drafted, debated, and passed in Portuguese. Tetum translations are rarely produced when the legislation is in draft form, and when translations are made they are usually by international actors (RDTL MOJ Official 2014). Legislators, the vast majority of whom do not speak Portuguese, are expected to debate the merits of legislation that they cannot understand. Most citizens continue to be bound by a legal system that is unintelligible absent translation. Even when the law is understood, legal interpretation is highly problematic because the original Portuguese law often assumes a radically different political, legal, and economic context. Language poses serious tensions in the basic core of the democratic process. Moreover, the entire law-and-justice apparatus still depends upon expatriate support from Lusophone countries (International NGO Manager 2014; RDTL MOJ Official 2014). Thus, the use of Portuguese will be a divisive issue for the foreseeable future. At the same time, Timor-Leste is not necessarily destined to this fate. Tetum offers a legal language almost all Timorese understand, and a legal system that truly embraces both Portuguese and Tetum could be far more accessible and equitable.

Overall Impact of Portuguese Bilateral Assistance

The impact of Portuguese bilateral assistance was uneven. A civil law system has been established and now operates nationwide. Portuguese ideals influence almost all laws and regulations. A functioning legal education regime was established at UNTL, the premier law faculty in the country, and the LTC. Both are overwhelmingly Lusophone. Portuguese became a marker of elite legal status. While some judicial proceedings are conducted in Tetum, Portuguese language has continued to be a prerequisite for LTC qualification. Moreover, *all* domestic judges, prosecutors, public defenders, and private attorneys are required by law to be LTC graduates (RDTL 2008). Although the societal penetration of Portuguese remains superficial, its influence over the legal profession and the formal state system is comprehensive.

Even minimal definitions of the rule of law require the law to be clear and accessible. Clarity is impossible when most parliamentarians cannot understand the laws they are voting on and only a tiny portion of citizens can understand the law that governs them. Unless there is a major change in either the level of understanding of Portuguese across society—or, alternatively, the legal system moves toward being truly bilingual—the state justice system will remain unable to fully uphold the rule of law.

Conclusion

The international community has much to be proud of regarding its work in Timor-Leste. While its efforts were by no means perfect, they were in general strategically sound and often beneficial. Programs addressed crucial justice sector needs by supporting the establishment, maintenance, and institutionalization of key organizations, such as the courts, prosecutors, and public defenders. International support helped to promote a rule-of-law culture with education and socialization initiatives as well as to expand access to the state system. Aid allowed more people to access the state legal system. International support also helped strengthen the links between state and non-state justice. All these efforts enjoyed state support, which helps explain why international assistance made such a significant contribution to advancing the rule of law.

International judicial state-building in Timor-Leste also exposed the limits of external assistance. Timorese state officials, on the whole, supported a democratic state underpinned by the rule of law, but it was a particular vision modeled on Portugal alongside an ideological commitment to reconciliation rather than retribution for past crimes. International initiatives incompatible with that vision fared poorly, most notably efforts to punish perpetrators of violence that risked undermining relations with Indonesia. Furthermore, state support does not guarantee positive long-term effects. Portuguese assistance enabled the creation of a Lusophone legal order, which has been envisioned by political elites ever since the struggle for independence. That legal order, however, threatens to become increasingly exclusionary as Lusophone legal elites entrench themselves while the percentage of Portuguese-speakers stays the same or even declines. International assistance has made institutions that underpin the rule of law possible and helped spread the cultural ideals necessary for it to take root. The lingering question is whether the state justice system can transcend its insularity and become a respected and accessible partner with the non-state system that still underpins legal order in most of the country.

6

From Competition to Combat

A Deeply Contested Legal Order in Afghanistan

Introduction

Shortly after the September 11, 2001 terrorist attacks, U.S.-backed Afghan resistance forces defeated the Taliban regime, which had been harboring Al-Qaeda. The new Islamic Republic of Afghanistan was soon established, with extensive international support. Even before its collapse, the Taliban regime had grown increasingly unpopular domestically and isolated internationally. An Afghan state that could provide not just law and order in culturally intelligible ways, along with accountability, democracy, and economic development, had broad appeal. Both domestic and international leaders pledged to make it a reality. These beliefs were embodied in the Bonn Agreement (Bonn Agreement 2001).

Given repeated declarations of support from the highest levels of government, widespread public approval, and major international investment, why did judicial state-building in Afghanistan achieve so little? This chapter argues that the new Karzai-led regime not only lacked a coherent strategy for advancing democratic governance and the rule of law in a highly competitive, legally pluralistic environment but actively opposed it. Consequently, powerful non-state judicial actors saw little benefit in engaging with the state for any reason other than short-term tactical gains and rent extraction. These law and governance failures also helped spur and sustain the Taliban insurgency.

Serious challenges were inevitable. Yet, a competitive legal environment does not preclude successful judicial state-building. Throughout history, almost every successful judicial state-building endeavor has required the state to suppress, outperform, or collaborate with non-state rivals (Fukuyama 2011). As in Timor-Leste, Afghanistan had a historical and cultural

Contending Orders. Geoffrey Swenson, Oxford University Press. © Oxford University Press 2022.
DOI: 10.1093/oso/9780197530429.003.0006

foundation for a legitimate state-based legal order. The fall of the Taliban regime was a golden opportunity for democratic state-building and progress toward the rule of law (Rashid 2008; Barfield 2010: 300–310; Jones 2010). That opportunity was squandered. Both domestic and international actors bear responsibility, but domestic actors had the preponderance of authority and power to shape the nascent political order. The Karzai regime focused on accumulating and retaining power while outmaneuvering potential rivals through patronage, international support, and increasingly dubious elections.

Summary of Arguments

Chapter 4 argued that domestic judicial state-building efforts in Timor-Leste achieved notable success because (1) despite fierce political competition, all major parties remained committed to constructing a democratic state underpinned by the rule of law; (2) there was a credible and sustained effort to develop institutions that promoted democratic accountability, inclusivity, and the rule of law; and (3) the state meaningfully engaged and collaborated with key non-state actors through successful jurisdictional division and local suco elections. This chapters illustrates how the leaders of the Islamic Republic of Afghanistan threw away the chance to build a more inclusive and effective democratic state that would be able to advance the rule of law. Not only did the regime lack a normative commitment to the rule of law, it systematically undermined institutional, legal, and political checks on its authority, including suppressing political parties, manipulating elections, and undermining judicial independence.[1] Moreover, the regime never seriously engaged with key tribal and religious non-state justice actors, who had long underpinned legitimate legal order in Afghanistan. The regime partnered with warlords, but it was a purely transactional relationship. The Afghan state also failed to mediate between the international community and key tribal and religious authorities. This divergent approach helps explain the failure of judicial state-building and the corresponding slide from competitive legal pluralism into combative legal pluralism with an increasingly potent Taliban insurgency.

[1] While the state's capacity was limited, the president faced very few constraints when determining policy or personnel.

Conventional Understandings of Judicial State-Building in Afghanistan Are Insufficient

Afghanistan is often cast as a land of anarchic tendencies (Tarzi 1993, Khalilzad 1997, Bearden 2001). The reality is rather different. Prior to the communist takeover, the country was quite stable. A legitimate state endured for centuries because state legal power was deeply, if volatilely, fused with tribal and religious authority. Only when the communists arose from outside the system did the country explode into conflict.

Likewise, the previous Taliban regime and its implications for judicial state-building in Afghanistan are widely misunderstood. Contrary to the claims of some prominent scholars (Rotberg 2002: 117; Suhrke 2011) and stated U.S. policy (Bush 2002: 4; USAID 2005), Afghanistan under the first Taliban regime was not a failing state, though it was certainly a brutal one.[2] Rather, the Taliban regime developed a legitimate law-and-governance system in a heavily armed, highly contentious society by drawing on the long-standing pillars of legitimacy in Afghanistan: cultural values, Islam, and the provision of public goods.

Afghanistan presented challenging but far from impossible terrain for establishing democratic governance and the rule of law. Yet, for many Afghanistan conjures up a sense of fatalism, which is distinct from but related to the idea that Afghanistan's natural position is anarchy. Unsurprisingly, as state-building faltered, this fatalism became ever more attractive and seemed ever more predestined. For some observers, it was there from the start. Referencing the "intensely war-like, independent, and anarchic traditions of many Afghan peoples," Ottaway and Lieven argued that democratic state-building was a "fantasy" given that previous state-building efforts relied on "sheer coercion" owing to the territory's geographic composition and the historically Pashtun-centric nature of the polity (Ottaway and Lieven 2002: 5, 2). Ironically, their solution of embracing regional warlords, albeit of some sort of gentler variety, was by and large what the international community ultimately did. It bears a striking resemblance to both the chaotic, discredited political order that originally spawned the Taliban as well as the culmination of nearly two decades of failed state-building efforts that sparked the Taliban's return to power in 2021.

[2] Other scholars have recognized that Afghanistan under the Taliban was not a failed state; see, e.g., Bøås and Jennings (2007) and Hehir (2007).

Fatalistic arguments, however, conflate what is difficult with what is impossible. They can neglect history and, more troublingly, serve to excuse poor policy choices and institutional arrangements. At first, the new regime enjoyed widespread popular support. A relatively stable, peaceful political order was certainly possible and indeed held for a number of years. The Taliban had previously established a monopoly on the use of force under even more chaotic conditions. Before that, former President Mohammad Najibullah (1987–1992) survived with far fewer resources and less popular goodwill than Karzai even at his lowest ebb. Before the Soviet invasion, Afghanistan had enjoyed decades of domestic tranquility. Nor is it right to think that the Afghan people are intrinsically prone to conflict. Afghan Pashtuns have certainly demonstrated a capacity for violence, but they also prize order, as evidenced by the paramount role of Pashtunwali and their widespread, centuries-long acceptance of a state, albeit a limited one.

The history of Afghan judicial state-building is a paradigmatic example of robust legal pluralism and an extraordinarily competitive legal environment. There are three enduring pillars of legal order in Afghanistan: state performance, culture, and Islam. While occasionally working in tandem, state, cultural, and religious authorities have also ruthlessly, and on occasion violently, competed over the right to determine how the population behaves. Competition intensified in the post-Bonn era as two new major clashes with state justice arose: competitive warlord justice and the Taliban's combative insurgent justice. In response, the Islamic Republic of Afghanistan engaged in numerous tactical alliances, which helped the regime endure but further compromised the regime's legitimacy and authority.

Chapter Overview

This chapter consists of two main parts. The first section traces Afghan judicial state-building from 1747 to 2001. State and non-state authorities never stopped competing, although the pillars of effective legal authority and governance—Islam, tribal culture, and provision of public goods—have remained remarkably stable. It starts by examining state law's long-standing competitors, tribal law and Islamic law, and explores how they have both competed and collaborated with state authority. It then examines the profound political and legal changes that occurred between 1978 and 2001. It concludes with the Taliban regime's rapid rise and swift collapse. The second part examines the post-conflict legal order envisioned by the Bonn Agreement and the 2004 constitution. It focuses on the role of the Karzai administration (2001–2014) in creating a corrupt, predatory state antagonistic toward democracy and the rule of law. The regime undermined electoral processes and ensured that Karzai's preferred successor, Ashraf Ghani, claimed the presidency despite widespread electoral fraud, which cast profound doubt on the legitimacy of the polls and substantial evidence Ghani may not have even secured the most votes (Mason 2014; Johnson 2018). The Karzai-led state also failed to capitalize on popular goodwill. Instead, it pursued political institutions ill-suited to a post-conflict state defined by competitive legal pluralism. These failures helped facilitate the rise of Taliban justice and an environment marked by combative legal pluralism.

Afghanistan from Origin to Soviet Invasion: The Pluralist Foundations of Legal Order

The origin of Afghanistan is conventionally dated to 1747, when a Pashtun tribal confederation elected Ahmad Shaw as ruler. From its onset, Afghanistan's political order was "dominated by the Afghan tribal-feudal socioeconomic framework" (Gregorian 1969: 47). The political order exemplified themes that persist to this day, including the frequent use of the Loya Jirga (grand council) as the foundation of governmental legitimacy, the prominence of Pashtuns,[3] the paramount role of Islam, a profound divide between the center and periphery, and a political order that relies on charismatic leadership. Before examining the historical trajectory of judicial state-building in Afghanistan, it is necessary to briefly discuss the major alternatives to secular state law: religious law and tribal law.

State Law as Religious Law

Until 1923, state law and Islamic law were synonymous. Shortly after its founding in 1747, the nascent state created "*Shari'a* courts in the urban

[3] At roughly 40 percent of the population, Pashtuns are the largest ethnic group and have been politically dominant since the emergence of an Afghan polity (Barfield 2010: 24). The other major ethnic groups are the Tajiks (30 percent of population), Hazaras (15 percent), and Uzbeks and Turkmen (10 percent) (Barfield 2010: 26–32).

centers" (Ghani 1978: 270). Courts operated based on religious law, and religious judges enjoyed substantial autonomy even as they furthered the reach of the state (Kakar 1979). State power was overtly religious. Kamali notes that "unlike tribalism, which compete[d] with the central government for political control, the religious establishment [was] basically pro-government" until the rise of the communist state (Kamali 1985: 8). Afghanistan has always been an overwhelmingly Islamic country. Currently, about 85 percent of the population are Sunni and 15 percent are Shia (Barfield 2010: 40). Historically, the dominant form of religion was Sunni Islam of the Hanafi School—long considered a moderate brand of Islam. It was based on a literate, professionalized religious jurisprudence that relied on trained clerics (*ulema*).

The official Islamic law system was deeply fused with state power, but its authority was concentrated in urban areas. Professional clerics often viewed "the rural customary system as illegitimate, particularly when they strayed from classical religious practices" and frequently attempted to leverage their "influence to demand the replacement of customary law practices with more standard *sharia* interpretations" (Barfield 2008: 352). Even after the introduction of secular legislation, Islamic values permeated nearly all state law.

Islamic law's universal claims were also rooted outside the tribal system (Roy 1990: 35). While less strict than the tribal codes, it is also less egalitarian and, in Pashtun areas, less culturally compelling. Islamic law mirrored the urban-rural divide in society. While urban religious elites enjoyed higher status, they could not control local religious authorities because Islam lacked a hierarchical leadership structure.[4] As Roy notes, "The *sharia* does not depend on any official body, church or clergy," and any decisions can later be reinterpreted (Roy 1994: 10). Distinctions in status among religious figures, therefore, largely hinged upon educational attainment or personal charisma rather than their position within a formal religious bureaucracy (Gregorian 1969; Roy 1990: 45–46). For example, village mullahs, though less educated, still projected authority and resolved local disputes based on their religious knowledge.

[4] In contrast, the world's largest Christian sect, Roman Catholicism, has a highly developed bureaucracy with a clear, hierarchical leadership structure and authorized representatives worldwide (Valuer 1971).

Tribal Law

Islamic law enjoyed the state's endorsement but was rivaled and often superseded by tribal law. Throughout Afghanistan's history, the most effective form of legal order in Pashtun areas was Pashtunwali, "an inclusive code of conduct guiding all aspects of Pashtun behaviour and often superseding the dictates of both Islam and the central government" (Carter and Connor 1989: 7). It was both "an ideology and a body of common law which has evolved its own sanctions and institutions" (Roy 1990: 35). Pashtunwali structured collective decision-making, military mobilization, conflict resolution, and social relations (Hager 1983: 83). Based on compensation and reconciliation, Pashtunwali's main tenets were straightforward:

(1) Honor is paramount, and the honor of women is to be protected.
(2) Gender boundaries must be rigidly maintained.
(3) One has a right to compensation, or *por*, when one is wronged.
(4) Revenge is tolerated and even encouraged.
(5) Apologies accompanied by *por* are to be accepted.
(6) Guests are to be sheltered.
(7) The *jirga* is to be obeyed. (Ginsburg 2011: 102)[5]

While sometimes characterized as informal law, it was in fact a highly formalized system that spelled out clear consequences for each violation. Pashtunwali offered a procedural and substantive legal framework underpinned by known rules and effective deterrents. There were often explicit compensation schedules for specific bodily or property harms. By sanctioning self-administered justice, it could be widely enforced even in the absence of official judicial authorities.

Pashtunwali provided a durable, effective, and legitimate legal order in a society with a weak central state, but it was a very severe system. Individual and collective enforcement of Pashtunwali to remedy violations was notoriously difficult to extract proportionately and frequently spurred further violence. Tribal law has few human-rights protections and was systematically biased against women (Kakar 2004). Because internal Pashtunwali obligations were paramount, the system frequently clashed with orders based on secular law or Islamic law and later with international law. Pashtunwali originated with

[5] For an alternative but similar statement of Pashtunwali, see Dupree (1973: 126).

the Pashtuns, but its ideology resonated throughout Afghanistan. The values of Pashtunwali "modified by local custom permeate in varying degrees all Afghan ethnic groups" and thus have long helped structure all non-state legal order in Afghanistan (Dupree 1973: 127).

Pashtunwali was, and is, applied through local gatherings known as *jirgas*. Non-Pashtuns used the *shura* process, which mirrored the operation of the jirgas (Glatzer 1998; Wardak 2003). While equity is stressed, the process is entirely male-dominated. Barfield, Nojumi, and Thier describe the jirga process:

> Everyone sits in a circle so that no one takes priority. All members have a right to speak and binding decisions are made by common consensus rather than voting. This may take considerable time (days, weeks or even months) or fail to come to a conclusion entirely. (Barfield, Nojumi et al. 2006: 9)

Jirgas could not sanction physical punishment, but they nevertheless fostered widespread compliance (Poullada 1973: 22; Kamali 1985: 4). The concept of a jirga was evoked at the tribal level for matters "central to the social order of the tribe," while Loya Jirgas were convened to address extraordinary important "vital national issues and make collective decisions" (Wardak 2003: 9, 13). All but one of Afghanistan's constitutions were ratified through Loya Jirgas, and jirgas and shuras remained the default dominant forms of dispute resolution unless resolution required the involvement of the state or another tribe (Figure 6.1; Carter and Connor 1989: 7; see also Wardak and Braithwaite 2013).

The State's Struggle to Establish a Secular Legal Order

From the Afghan state's creation, it was defined by competitive legal pluralism.[6] Roy contends that "the history of the Afghan state (*dawlat*) from 1747 to the present is bound up with the search on the part of the state

[6] "Secular" is used here in a procedural and an administrative sense, not a substantive one. It refers to the origin of the law from state political authorities rather than religious authorities, not its normative content. Afghan state law was almost always explicitly religious in content and values. All Afghan constitutions, with the exception of the 1980 charter, required consistency with Islam and stressed the importance of Islam to Afghan political life.

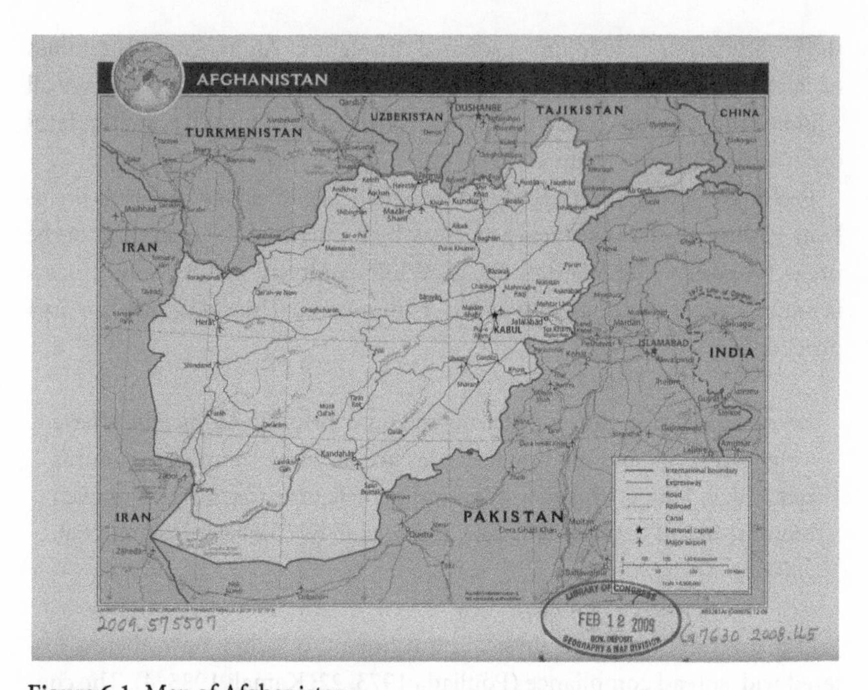

Figure 6.1 Map of Afghanistan

Credit: Central Intelligence Agency 2008, Library of Congress Map Collection, https://www.loc.gov/item/2009575509/

bureaucracy for autonomy from the tribes" (Roy 1990: 4). Since the constitutional era began in 1923, the Afghan state has searched for sources of legitimacy that depend less on tribal support, external religious sanction, or individual charismatic leadership. This section traces that struggle from its halting progress through the collapse of constitutionalism into full-scale civil war in 1979.

Conceptualizing Constitutionalism

The major constitutional frameworks shaped the judicial state-building attempts and engagements with non-state legal orders. While rulers never followed their constitutions to the letter, they possessed immense symbolic value and projected each regime's aspirations domestically and internationally (Alexander 2001). The constitution is "an essential symbol of the modern state" as well as of states seeking to modernize (Elkins, Ginsburg et al. 2009: 42). Not coincidentally, Afghanistan's first constitution was drafted in 1923, shortly after the country achieved full independence from British influence in 1919, and underpinned an ambitious modernization program (Poullada 1973). Afghan constitutions have always fallen short of what Sartori dubbed a "proper," fully actualized constitution that structures and constrains political power (Sartori 1962: 861). Proper constitutions are contrasted with "nominal constitutions," which "describe a system of limitless, unchecked power" and "façade constitutions" that do not illuminate how the state works or limit its power (Sartori 1962: 861). Still, constitutions can bolster legitimacy even in authoritarian regimes (Ginsburg and Moustafa 2008a, b).

In a society with a high degree of legal pluralism society and a weak state, the constitution provides key insights into the fluid relationship between state and non-state legal orders. Constitutions matter immensely to state leaders as well as to tribal and religious elites. State, tribal, and religious authorities have consistently fought over the constitutional scope of state authority, sometimes even resorting to violence and rebellion. Yet, it is important not to reduce political analysis solely to legal analysis. The most fundamental struggles in societies with high degrees of legal pluralism, and especially combative or competitive legally pluralist ones, are over not the content or interpretation of state law but whether the state can co-opt or marginalize the "families, clans, multinational corporations, domestic enterprises, tribes, patron-client dyads—which make rules against the wishes and goals of state leaders" (Migdal 1988: 31).

The Dawn of Constitutionalism: Reform and Backlash

Prior to the reign of King Amanullah (1919–1929), all law was explicitly either tribal or Islamic. In 1923 Afghanistan's first constitution was promulgated. The foundational law and other secular legislation aimed to transform both the state and society. The constitution proclaimed Afghanistan an independent, unitary monarchy bound by "Islamic law and this fundamental code" (Kingdom of Afghanistan 1923: Art. 1, 4). An independent judiciary would apply the law uniformly nationwide (Arts. 51, 53) and make decisions in accordance with "Islamic law and the principles of civil and criminal courts" (Art. 21).[7] Secular civil, commercial, and penal codes were

[7] While compliance with Islam may seem to dramatically limit state power, Nadir Shah and his dynasty merged Islam and the state-building endeavor through the notion that "to be a good Muslim

promulgated (Poullada 1973). The nascent constitutional legal order drew inspiration from Islam but was not bound by the clerical establishment's views. Amanullah's reforms generated deep suspicion from two pillars of monarchical rule: the tribes and the religious establishment. A major revolt followed in 1928, instigated by powerful Pashtun tribal interests underpinned by the moral authority lent by the religious establishment. The regime soon collapsed. In 1930 a new dynasty emerged, known as the Musahibans, founded by Muhammad Nadir Shah.

The 1931 Constitution and Tentative Judicial State-Building

Nadir Shah's regime attempted to legitimize itself through a constitution. The 1931 constitution reflected the prominence of tribal authorities and the religious establishment. Article 1 stated that "the faith of Afghanistan is the sacred faith of Islam," specifically the Hanafi school, and required the king to be an adherent (Kingdom of Afghanistan 1931: Art. 1). Sharia was the basis of all law, and all legislation must be consistent with it (Art. 65). The constitution promised formal legal equality and protected certain legal rights, but those rights and duties existed "under the Sharia law and the law of the state" (Kingdom of Afghanistan 1931: Art. 13). It also established the first official representative assemblies (Arts. 27–72). The constitution entrenched tribal authority. Pashtun tribal groups were "exempt[] from military conscription and taxation" (Shahrani 1986: 52). Loya Jirgas were limited to every three years, and any constitutional reform or other major policy change required the consent of this tribal-, particularly Pashtun-, dominated institution (Gregorian 1969: 305).

While the 1931 constitution constrained state authority relative to religious and tribal power, it helped solidify the state authority that remained. As Gregorian argues, the 1931 constitution "institutionalized the power of religious establishment" but also "permit[ted] gradual reforms of the judicial system in keeping with the requirements of a modern state" (Gregorian 1969: 340). Nadir Shah was assassinated in 1933. When Mohammad Zhair

and a good Afghan one had to contribute to the stability, prosperity, and progress of Afghanistan under the guidance of the monarchy" (Gregorian 1969: 292–293). The MOJ even supervised implementation of Islam (Gregorian 1969: 295). In other words, the logic was circular: as good Muslims ran Afghanistan, their actions by definition were consistent with Islam.

Shah assumed the throne, the limited, tentative process of judicial state-building nevertheless continued.

The 1964 Constitution and Tentative Liberalization (1964–1973)

The early 1960s saw the culmination of a prolonged internal power struggle between the king and his cousin, Prime Minister Sardar Mohammad Daoud. In 1964 the king attempted to solidify his position through a more inclusive and more secular constitution (Tarzi 2012: 220), while still emphasizing the monarchy's Islamic credentials as a source of legitimacy (Shahrani 1986: 56). Islam remained "the sacred religion of Afghanistan," and legislation had to be consistent with it (Kingdom of Afghanistan 1964: Art. 2, 64). Islamic law remained the default rule absent state regulation (Art. 69).

The 1964 constitution reconceptualized state, tribal, and religious authority. Constitutional provisions for judicial independence (Constitution of Afghanistan 1964: Art. 97) were coupled with meaningful attempts to promote it (Kamali 1985). Under Article 88, state judicial authorities claimed exclusive jurisdiction to resolve all legal disputes according to state law. Ultimately, the 1964 constitution and related reforms continued the trends of legal codification and an increasing preference for law faculty graduates over Sharia graduates, while steadily decreasing the religious establishment's power (Dorronsoro 2005: 57; see also Edwards 2002).

Outside the legal sector, the 1964 constitution facilitated broader political participation. Zahir Shah, however, never fully embraced multiparty democracy. The press still faced significant restrictions. Parliament never directed state policy or checked royal authority (Barfield 2010: 211). Political liberalization facilitated greater popular mobilization but not, ultimately, democratization. Marxist-inspired groups and Islamists both lacked a commitment to representative government.[8] Amid growing political tensions, economic malaise, and a devastating famine, Daoud seized power

[8] While not formally political parties, they fall squarely within the phenomenon aptly described by Gunther and Diamond as "anti-system" parties. These groups favor replacing the existing order "with a regime that would be more uniformly committed to the achievement of their programmatic objectives" (Gunther and Diamond 2003: 171).

in a coup in 1973. He immediately disbanded the monarchy and declared Afghanistan a republic.

Daoud's Non-Constitutional Regime (1973–1978)

Daoud's seizure of power was unprecedented in Afghanistan's history because it did not depend on tribal or clerical support. Technically, Daoud's republic initially retained the overarching legal framework but quashed movement toward judicial independence (Kamali 1985: 240–243). Still, the government continued legal reforms, such as enacting comprehensive penal and civil codes (Kamali 1985: 240–243). Daoud's regime emphasized "centralisation of power, increased state regulation of the economy, and a whole cluster of social reforms, including equal rights for men, women and national minorities" (Saikal 2004: 176). The end of state-sanctioned favoritism of Pashtun tribes and the religious establishment further undermined the traditional pillars of state legitimacy. In 1978 the People's Democratic Party of Afghanistan (PDPA) toppled Daoud in a coup. The PDPA, now led by President Noor Muhammad Tarki, declared the "Democratic Republic of Afghanistan."

Non-State Order from 1923 to 1978

Throughout this period the dominant legal order continued to be non-state justice, despite the state's significant gains in authority, capability, and resources. Goals were modest: the state simply demanded the allegiance of tribal leaders and the local population. Local politics and dispute resolution, however, was generally left to local populations. Competing political and legal authorities clashed, but these tensions did not threaten the regime. Urban areas, in contrast, proved an existential threat. New groups, notably communists and Islamists, sought to dominate the state rather than avoid it. When communists succeeded in seizing control of the state, Afghanistan soon exploded in open conflict between a primarily rural religious-tribal non-state order and urban communist regimes. This violence largely annihilated the legal and political state structures that had been built over the last half century.

The relationship between the state legal order and tribal and religious orders maintained an unsteady truce. It was highly competitive but rarely threatened the overarching legal and political order by becoming combative. The state had the preponderance of power, but it required at least tacit tribal consent (Shahrani 1984), and it largely deferred local policy to local people. In turn, rural areas sought autonomy but not independence. Tribal leaders appreciated the state's provision of services they could not.

The state leaned heavily on Islam for legitimacy (Shahrani 1986: 56). Government subsidies curried favor with the religious establishment while simultaneously asserting greater control through increased regulation (Canfield 1986). Professionalization weakened the authority of non-state religious actors relative to the state. Religious ideology proved much more difficult to tame than religious leaders. Islam's decentralized structure, which made collective action difficult, would also provide major political openings for the Islamist resistance fighters and later for the Taliban.

The Collapse of Constitutionalism and the Rise of Violent Conflict

While Afghanistan's state capacity remained limited, "considerable progress towards the achievement of a national state had been achieved prior to the Marxist seizure of power in 1978" (Newell 1986: 110). The old regime could project effective force when local powerbrokers threatened to challenge the state's authority (Barfield 2010: 224). After the communist takeover, the space for violent political entrepreneurs increased dramatically, allowing the emergence of warlords as yet another major competitor for authority.

The communist regime initially pursued an unabashedly radical leftist agenda first under Tarki and then Hafizullah Amin. In the beginning the government did not seek support from tribal leaders or religious authorities. However, the results were disastrous, and the regime found itself in a precarious position. The Soviet Union invaded in 1979, an intervention that sought to remove the unpopular Amin, unify Afghanistan's competing communist factions, and "frighten or deter the rebels" to provide the new President, Babrak Kamal, time "to rebuild an army and re-establish control over the countryside" (Collins 1986: 77). The invasion violated international law and prompted major international support for the anticommunist resistance dubbed the mujahedeen, particularly from the United States and Pakistan. Afghanistan was soon engulfed in a major conflict.

Legal Order under Communist Rule (1980–1986)

Communism bolstered state legitimacy in many developing countries worldwide (Huntington 2006), but it was deeply unpopular in Afghanistan (Kakar 1995). Lacking the legitimacy of previous governments, the regime leaned heavily on coercion (Maley and Saikal 1992). The communist regime did nevertheless attempt to craft a legalist veneer. In terms of organization, the judicial system remained largely unchanged (Kamali 1985: 242). Substantively, "The Marxist government attempted to introduce a Soviet style judicial system, but these changes were rejected before they took root" (Wardak 2004: 319). The Karmal regime promulgated a provisional constitution in 1980 (Democratic Republic of Afghanistan 1980). The constitution made half-hearted attempts to establish Islamic legitimacy by placing Islamist symbols on the flag and recognizing Islam as the state religion (ibid.: Arts. 64–65). In practice, the constitution was a "facade" constitution, which neither limited state power nor offered "reliable information about the real governmental processes" (Sartori 1962: 861). The collapse of constitutionalism coincided with the rapid erosion in state legitimacy and administrative capacity.

A Communist-Islamist Order (1986–1992)

Upon assuming the presidency with Soviet support in 1986, President Najibullah jettisoned the regime's communist ideology, promoted national reconciliation, and reestablished links with religious leaders and tribal authorities by emphasizing his Pashtun heritage. In other words, "The communist state began to play the traditional tribal game" (Roy 1994: 150). A new constitution, modeled on the 1964 constitution, was promulgated in 1987, but it did not meaningfully constrain state power. Rather, its purpose was symbolic. It emphasized Islamic values and greater inclusivity (Democratic Republic of Afghanistan 1987: Art. 2). Ultimately, it did not bolster the regime, nor did it achieve Islamic legitimacy (Tarzi 2012: 230). In 1988 the Soviet Union signed an agreement with Pakistan, co-guaranteed by the United States, to remove their forces by February 1989 (Jones 1990: 191–192). Unexpectedly, Najibullah retained power until 1992, when the Soviet Union itself dissolved and subsidies ceased.

The Civil War's Impact on Non-State Order

The nexus of religious and tribal power underpinned mujahedeen resistance to Marxist rule. Ideologically, Islam sustained resistance efforts (Roy 1990). Sinno's detailed analysis of the resistance's structure argues that '"the Afghans' attachment to Islam and their code of honor—was evidentially a highly effective motivational force" as well as "a necessary condition for resilience," if not a sufficient one (Sinno 2008: 134). The long-standing non-state legal structures continued to structure local legal order, albeit it often with new leadership (Roy 1994: 164). Non-state justice still regulated life for ordinary people, but many resistance leaders acted with increasing impunity (Dorronsoro 2005: 210).

At the same time, the Soviet Union became increasingly reliant on militias (Giustozzi 2012a: 53–56). The civil war thus empowered a new class of non-state justice providers with an inherently competitive outlook toward state authority: warlords.[9] Warlords engaged with the state, but only when compelled or if it was strategically advantageous to do so (Mukhopadhyay 2014). They were structurally opposed to the judicial state-building process (Reno 1998, Marten 2006), and even when working with the state they rejected attempts to fully implement state law in their territory. Warlords demanded either exemptions from the state legal order or sought to implement their "own (often arbitrary) code of behaviour" (Ahram and King 2012: 172). They established substantial authority, underpinned by force, patronage, and charisma, over significant parcels of territory and capitalized on the space for autonomy created by the communist government's dependence on militia and the highly factionalized resistance.

[9] I follow Ahram and King's definition of warlords as "armed agents who wield some degree of civil power and claim some kind of local sovereignty over a defined region while paying allegiance to one or more stronger powers" (Ahram and King 2012: 172). This definition underscores the role of force in maintaining warlords' authority as well as their ability to provide some non-state order while simultaneously avoiding normative judgments. However, warlords' deference to higher authorities is often largely theoretical.

Clashing Islamist Orders (1992–2001)

Mujahedeen Fiefdoms (1992–1996)

In 1992 the mujahedeen resistance triumphed. While religion helped unify the resistance, the resistance factions lacked a shared vision for a post-communist order. In April 1992 they proclaimed the Islamic State of Afghanistan, which existed until September 1996. As Rubin wryly notes, "Perhaps this entity was Islamic, but it was hardly a state, and it certainly did not rule Afghanistan" (Rubin 2002: 272). The state suffered from chronic legitimacy and economic problems as well as active insurgency from the remaining warlords. The state legal system was minimal and chaotic. Each regime faction "appointed their members to the judiciary, most of whom had no legal education or even professional work experience," which further undermined the state's legitimacy (Hamidi and Jayakody 2015: 20). The situation quickly deteriorated into civil war. Legal order became overwhelmingly non-state and highly localized, resulting in a harsh, violent, and often arbitrary brand of combative legal pluralism.

The Taliban's Brutal Legal Monopoly (1996–2001)

Prior to the emergence of the Taliban, "the country was divided into warlord fiefdoms and all the warlords had fought, switched sides and fought again in a bewildering array of alliances, betrayals and bloodshed" (Rashid 2001: 21). Against this backdrop the Taliban transformed from a student group in early 1994 to a major military force in less than a year. The Taliban gained control of Kandahar and quickly established order, gaining a reputation for "religious fervor and purity" (Marsden 1998: 46). By the end of 1996 the Taliban exercised sovereignty over the capital along with roughly 75 percent of the country (Nojumi 2002: 159).

The Taliban's military success translated into a monopoly on the legitimate use of force (Rubin 2000). Many victories came without a significant military engagement, and no viable insurgencies occurred within Taliban controlled territory. The Taliban capitalized on long-standing pillars of legitimacy. An overwhelmingly Pashtun movement, the Taliban exploited the widely held belief that only Pashtun rulers were legitimate (Goodson 2001: 125) and advocated an interpretation of Sunni Islam relatable to rural areas (Edwards 2002: 294). Since the birth of Afghanistan, Islamic credentials have been vital for effective rule. Yet, the Taliban were something new. Their harsh brand of Islam and commitment to uniformly imposing their belief system nationwide diverged from Afghanistan's long history of religious tolerance toward different forms of Islam (Rashid 2001: 82). The Taliban merged religious and state authority, and warlords were defeated or displaced. In an environment of extraordinarily combative legal pluralism, the Taliban had established a legal monopoly.

The Taliban regime was fundamentally legalistic. It pledged "to bring peace to Afghanistan, establish law and order, disarm the population, and impose sharia" (Rashid 1999: 24), focusing on "social stability" and "achiev[ing] law and order in society through Islamic ways" (Gohari 2000: 55). Over time, however, the Taliban regime developed greater state capacity focused on "security, justice, and the observation of religious regulations" (Dorronsoro 2005: 278). Nevertheless, the court system featured very little institutionalization (Jones, Wilson et al. 2005: 68), but promulgated ad hoc edicts as it deemed necessary. Procedurally and substantively, Taliban law violated international human-rights norms (Franck 2001), and its treatment of women was particularly heinous (Moghadam 2002).

The Taliban claimed that its legal system was purely Islamic, but it also reflected Pashtun tribal values. Pashtunwali and other non-state adjudication systems were largely tolerated and not seen as inconsistent with Taliban understandings of Islam. Sinno goes so far as to posit that the reason that the Taliban refused to turn over Osama bin Laden was that it would violate Pashtunwali's requirements of hospitality toward guests (Sinno 2008: 244). Tribal non-state justice mechanisms continued to resolve most disputes, although the Taliban occasionally influenced outcomes.

Ultimately, the Taliban regime collapsed owing to a major foreign policy miscalculation rather than domestic unpopularity. The Taliban harbored Al-Qaeda members during the planning and execution of the September 11, 2001 terrorist attacks. Shortly thereafter the United States backed the anti-Taliban Northern Alliance in a major anti-Taliban offensive. By December 2001 the Taliban regime had collapsed under the military onslaught, and its leaders fled.

Judicial State-Building and Legal Pluralism under the New Regime (2001–2014)

From Transitional Governance to Constitutional Order (2001–2004)

While international intervention was necessary to dislodge the Taliban regime, it was already deeply unpopular by 2001. The country was ready for a fresh start, one that promised legitimate law and order as well as democratic accountability. While the new state explicitly endorsed the goals of liberal democratic state-building, democracy and the rule of law remained elusive. Failure, however, was not predestined. The policy decisions undertaken by the state officials and other domestic actors, most notably the regime of Hamid Karzai, played an important role in the outcome. Not only was the judiciary prevented from being professionalized and independent, but it was also incentivized to become predatory. The executive was overpowered, unaccountable, and unrepresentative. Political parties were undermined and suppressed. The election system was badly flawed and unrepresentative. Elections themselves lacked integrity. In short, the Karzai regime actively prevented progress toward the rule of law. Moreover, it never seriously attempted to engage the non-state justice actors rooted in tribal and religious authority that had long been the foundation of legitimate legal order in Afghanistan.

The Bonn Agreement

The Bonn Agreement set the trajectory for state reconstruction and efforts to establish the rule of law. The agreement established a thirty-person, six-month interim government headed by Karzai. It reinstated the 1964 constitution, absent the monarchy, and existing legislation not inconsistent with that document (Bonn Agreement 2001: Sec. II.1). Judicial authority was vested in the Supreme Court (Sec. II.2), while a new institution, the Judicial Commission, was given the tall order of "rebuild[ing] the domestic justice system in accordance with Islamic principles, international standards, the rule of law and Afghan legal traditions" (Bonn Agreement 2001: Sec. II.2). Discussions omitted non-state justice. The agreement called for disbanding all non-state military forces but lacked enforcement provisions. In fact, the interim government increasingly depended upon the militias to project "state" authority outside the capital. Bonn also excluded the Taliban from the reconstruction process and laid the foundation for combative legal pluralism between the state and the reconstructed Taliban.

Emergency Loya Jirga (2002)

The roughly fifteen-hundred-person Emergency Loya Jirga in 2002 authorized the transitional regime until the constitutional drafting process was completed and elections were held. Karzai remained chief executive and the dominant decision maker. While the international community embraced the outcome, domestic perception was more mixed (Wardak 2003). The U.S. influence on the process was clear, as the U.S. envoy "strong-armed the king into throwing his support behind Karzai" (Barfield 2010: 297). Warlords, who quickly remerged after the Taliban fell, exerted influence through "political pressure, intimidation, and [the] spread of money" (Jalali 2003: 177). The Emergency Loya Jirga allowed the state-building process to continue but did little to promote consensus, the rule of law, or legitimate governance, or to establish a constructive relationship with key non-state justice actors.

The New Regime (2004–2014)

Karzai and his allies stressed the need for a strong executive to bring order, peace, and stability. Before the 2004 constitution was enacted, Karzai declared, "In countries where there are no strong institutions, where the remnants of conflict are still there, we need a system with one centrality, not many centres of power" (Karzai, quoted in International Crisis Group 2003: 3). Yet, it was never clear how centralization would help consolidate democracy or even establish a nondemocratic but nevertheless legitimate order.

The Karzai regime further centralized power over the next decade by undermining constitutional checks and balances through legal and electoral manipulation as well as through patronage and the application of state power. In theory, the new state institutions could support democratic governance bound by the rule of law, and the regime tolerated political competition. In practice, the regime lacked a normative commitment to both democracy and the rule of law. Centralization did not provide the promised order; rather, the state's ability to deliver public goods deteriorated. Likewise, the state's authority came increasingly to be contested as the insurgency became grew stronger.

The new state largely eschewed the traditional pillars of legitimate authority. Religion was acknowledged, but the state was not clothed in religious legitimacy. Karzai himself lacked any robust religious credentials. Having recently dislodged the Taliban, the regime's international supporters were wary of the new state's being too closely linked with Islamic ideals and emphasized the promotion of human rights, particularly women's rights. Thus, international priorities clashed with Pashtunwali and Islamic law as understood by many in Afghanistan, which the Taliban then used to criticize the regime. The state's tribal relationships were also often tense. Consequently, the state depended largely on the potential for legitimacy offered by elections and provision of public goods, most notably a just legal order and economic development. The Islamic Republic of Afghanistan's institutional structure was equally problematic, but if it had performed well and improved the quality of life, popular support certainly would have been greater. As with its 1964 predecessor, the new constitution increased political participation but failed to promote a democratic order that embraced rule-of-law ideals.

The post-Bonn regime sought to consolidate and retain power rather than build state capacity. Although the state promulgated various strategic plans backed by international donors (for example, see London Conference on Afghanistan 2006; Islamic Republic of Afghanistan 2008, 2010), the regime was not committed to any of them (Suhrke 2008). Instead, the regime behaved like a "vertically integrated criminal organization . . . whose core activity was not in fact exercising the functions of a state but rather extracting resources for personal gain" (Chayes 2015: 62). There was no road map for how the state would realistically improve its provision of public goods or make life better for ordinary people. As Chapter 7 will highlight, the international community often enabled, and occasionally encouraged, the regime's destructive behavior. Nevertheless, the state and its failings inescapably reflected domestic politics.

The 2004 Constitution

In December 2003 a Loya Jirga ratified a new constitution that envisioned a strong, highly centralized state. The constitution strove for religious legitimacy, proclaiming, "Islam is the religion of the Islamic Republic of Afghanistan" (Islamic Republic of Afghanistan 2004a: Art. 2). Religious law provided the default rules absent relevant state legislation, and a supremacy clause stipulates that "no law shall contravene the tenets and provisions of the holy religion of Islam in Afghanistan" (Art. 3). The relationship of state justice to its other major competitor, tribal law, was not mentioned, though state law was constitutionally established as superior.

Executive Branch

Executive power was fully vested in a strong, directly elected president who is limited to two terms (Arts. 61–62). There was no prime minister. The president appointed cabinet officials, legal personnel, and regional governors (Art. 64). The president enjoyed veto power, pardon authority, the ability to declare war subject to the National Assembly's approval, and numerous other executive functions as both the head of state and head of the government (ibid.).[10]

Executive authority also impinged on and undercut legislative authority. Only the executive could propose financial and budgetary legislation (Art. 95). Presidents of the Islamic Republic of Afghanistan could legislate without a clear delegation of legislative authority under a clause allowing the executive to "devise as well as approve regulations," provided they were not "contrary to the body or spirit of existing law" (Art. 76). The president could fill legislative gaps (or supposed gaps), which allowed the executive to circumvent the legislature through decree laws even in matters where the parliament clearly has spoken. Under Article 79, when the lower house (Wolsei Jirga) was in recess, "the government shall, in cases of immediate need, issue legislative decrees," provided they do not address financial or budgetary matters.[11]

Presidentialism does not necessarily preclude successful state-building or democratization or the rule of law (Mainwaring and Shugart 1997). Robust authority in state institutions "allow[s] the community to deploy that power to enforce laws, keep the peace, defend itself against outside enemies, and provide necessary public goods" (Fukuyama 2014: 24). However, insufficiently checked state power invites authoritarian behavior. The rule of law helps structure state authority and ensure that it is used appropriately. The

[10] Constitutional accountability mechanisms were minimal and cumbersome. Neither the president nor members of his government could be tried in regular courts (Art. 69, 78). Removal of the president required an accusation backed by one-third of the Wolsei Jirga. If two-thirds of that body agreed, a Loya Jirga would be established. If two-thirds of the Loya Jirga agreed with the charge, then a special court would be established to hear the charges.

[11] Once in session, the legislature had thirty days to reject the law, a tight time frame that requires substantial collective action.

nascent Afghan state faced real challenges and needed authority, but its exercise of power was inconsistent with the rule of law, undermined the state's legitimacy, and helped spur the insurgency (Chayes 2015).

A presidential system concentrates authority, which makes it at least theoretically easier to exercise. But by concentrating power, presidentialism "introduces a strong element of zero-sum game into democratic politics with rules that tend towards a 'winner-take-all' outcome" (Linz 1994: 18). This dynamic can be particularly explosive in deeply divided post–civil conflict settings (Horowitz 2000), as violence almost inevitably deepens ethnic divisions, even ones that were not particularly salient before the conflict (Kalyvas 2006).

Legislative Branch

While the president was the dominant political actor, the bicameral National Assembly possessed some real authority. The directly elected lower house, the Wolesi Jirga (Art. 91), had to confirm all major presidential appointments. The upper house, the Meshrano Jirga, consisted of provincial, district, and presidential appointees. The National Assembly was not shy about using its power, however. It repeatedly rejected presidential nominees and government-backed legislation, demanded investigations, and called for the resignations of cabinet members and court personnel. Imperfect as it was, the legislature was the only state institution that meaningfully challenged presidential authority within the bounds of constitutional structures. In practice, however, parliamentary effectiveness was limited by corruption, weak political parties, a convoluted electoral system, and, again, a lack of commitment to democracy and the rule of law.

The Judicial Branch

Rooted in the civil law tradition, state legal order was based on codes and legislation. Judicial independence was enshrined in the constitution (Art. 116). Judicial power was largely vested in a nine-person Supreme Court, appointed by the president and endorsed by the Wolsei Jirga (Arts. 116–117). Judicial jurisdiction was comprehensive and could not be removed by legislation, only by constitutional amendment (Arts. 120, 122). The Supreme Court possessed the right of judicial review for "laws, legislative decrees, international treaties as well as international covenants" (Art. 121), but the constitution did not explicitly mention the right to *constitutional* review.[12] Furthermore,

[12] The textual ambiguity surrounding constitutional review by the Supreme Court or an independent commission for supervision of the implementation of the constitution sparked immense controversy (Kamali 2014).

judicial review was structurally limited because standing to challenge a law requires referral by either the government or the courts. Although the judiciary proposed lower-court judges, presidential approval was required (Islamic Republic of Afghanistan 2005). The president had tremendous authority to ensure that only sympathetic judges ever reached the bench. The president could mitigate or pardon criminal offenses and even retire sitting judges (Art. 64). Consequently, judicial power, particularly in relation to the executive, was structurally precarious.

Elections and the Nascent State Political Order

In democratic states, elections structure the state-sanctioned political order. Elections can help legitimize the emerging political authority, promote accountability, allow the population to guide state policy, and secure support from potential spoilers. At the same time, even free and fair elections can spark conflict and deepen divisions. The election system itself is vital, because "an inappropriate or flawed electoral system can retard democracy's progress as much as warlords, religious fundamentalists, and corrupt business leaders combined" (Reynolds 2006: 105). Local elections can also make non-state justice actors more open to collaboration with the state, as in Timor-Leste. Election systems were particularly important in Afghanistan because there was no natural party of government stemming from a successful independence or revolutionary movement.[13] Vibrant, credible electoral competition helps consolidate the rule of law by promoting governmental accountability and ensuring the judiciary does not become the tool of one particular group (Lipset 2000).

Presidential Elections

Afghanistan's first democratic election of a head of state was held in October 2004. Eight candidates ran for office. Roughly 10.5 million Afghans registered to vote, with 80 percent of eligible voters casting ballots. Karzai emerged victorious, with over 55 percent of the vote. Despite security concerns and some irregularities (Coburn and Larson 2014: 53–55), Karzai's election enjoyed legitimacy both domestically and internationally (Goodson 2005). The 2004 election was a rare bright spot in a state-building process.

[13] Leaders of the victorious Northern Alliance were largely from minority groups and thus were not palatable to the Pashtun plurality.

President Karzai retained power in the 2009 elections. This election, however, further undermined state authority rather than bolstering it (Suhrke 2011: 153). The voter registration process was tainted by irregularities, and the election itself was marred by large-scale fraud (Worden 2010). While the international community was initially reluctant to acknowledge the extensive fraud, pressure grew, and Karzai accepted a runoff. His opponent, Abdullah Abdullah, withdrew before the vote actually took place. Electoral fraud increased further in the 2014 elections, in which Karzai heavily manipulated the election results in favor of Ashraf Ghani (Shahrani 2015; Johnson 2018).

Political Parties

Political parties are essential to establishing, consolidating, and sustaining democracy. In Diamond's words, "Political parties remain an indispensable institutional framework for representation and governance in a democracy" (Diamond 1999: 96). Parties facilitate the representation and mobilization of diverse societal interests in state institutions. This role is particularly important in places like Afghanistan, where potentially anti-system actors, such as warlords, enjoy electoral advantages thanks to their influence over local populations. While political parties are not inherently good or virtuous, they are a prerequisite for any "serious attempt at democracy" (Carothers 2006a: 10–11), as they serve as a check on executive authority and structure collective public choice.

The establishment of robust political parties in Afghanistan presented serious challenges. Many political parties that remerged after the collapse of the Taliban regime had been discredited during mujahedeen rule (Bhatia 2007). New parties were only weakly rooted in society. The overarching structure of party competition reflected choices made by the Karzai regime (Reynolds 2006, Coburn and Larson 2014). U.S. Ambassador Zalmay Khalilzad likewise favored a strong presidential system, but the final decision was made by Karzai's regime (Suhrke 2011). Karzai sought "to prevent the formation of strong national political parties and keep the influence of organized political factions within the parliament limited" (Giustozzi 2013: 328).

The Political Parties Law and Its Consequences

The Karzai administration undertook a concerted campaign to undermine political parties. The initial 2003 political parties law promoted the proliferation of weak political parties rather than strong, broad-based ones

(Transitional Islamic State of Afghanistan 2003: Art. 8–9; International Crisis Group 2005). Under the revised political party law of 2009, parties needed ten thousand individual members, verified by identification cards (Islamic Republic of Afghanistan 2009). This made the process of securing legal party status quite difficult. Only five parties successfully registered prior to the 2010 parliamentary elections, "which meant only 31 of approximately 2,500 candidates could advertise a party ticket" (International Crisis Group 2013: 5).

The party regulation tightened further in Karzai's second term. Under legislative changes in April 2012, parties needed active offices in at least twenty of Afghanistan's thirty-four provinces (Ministry of Justice 2012). The existence of these offices was aggressively policed. State officials even admitted that the regulation was aimed at limiting political parties (International Crisis Group 2013: 7). When the April 2013 deadline arrived, not one party had met the stringent requirements (Ariana News 2013). Although some parties continued to operate "informally" (U.S. State Department 2015a: 30), their continued existence was left at the whim of the state.

The Election Law and its Consequences

In 2004 the election law was promulgated. The official electoral study commission strongly recommended proportional representation. Instead, Karzai's administration selected the rarely used single non-transferable vote (SNTV), despite serious concerns from international advisers (Reynolds 2006). Under SNTV, districts have multiple seats, and the top vote recipients in each district receive seats. Each province had their total number of representatives allocated by population, with a total of 249 seats nationwide. The representatives allocated to each province ranged from two seats in Panjsher to thirty-three in Kabul. SNTV is a "semi-proportional" electoral system that allocates seats based on votes with a method "somewhere between the proportionality of PR and the majoritarianism of plural-majority systems" (Reilly 2001: 17). Coburn and Lawson highlight:

> SNTV served a number of high-level Afghan political interests while creating chaos for local communities of which many local leaders would take advantage. SNTV made the process of choosing a candidate to vote for simple perhaps but it made the outcomes incredibly complex and easy to manipulate from above. Without any requirements of party affiliation, it discouraged organized opposition and increased the likelihood of a

discouraged legislature that would not be able to counter executive decrees or decisions effectively (Coburn and Lawson 2014: 58).

The 2005 Parliamentary Elections

After a delay of more than a year, the Wolesi Jirga and provincial council elections were held in September 2005. This marked the formal end of the transitional process initiated in 2001. Approximately 6.4 million people voted. Wilder notes that "about 50 percent of registered voters [participated], a sharp decline from the 70 percent that voted in the Presidential elections a year earlier," a decline he attributed to widespread popular dissatisfaction with Karzai's government (Wilder 2005: 32).

Parliamentary elections posed serious administrative, logistical, and security issues. Party affiliations were not required for parliamentary candidates, and even when applicable were not listed on the ballot. Given the large number of candidates and parties, ballots were sometimes many pages long. Consequently, in the 2005 election voters could not support candidates based on their political party without advance knowledge.

Even in established democracies, SNTV disfavors large, broad-based parties and promotes party fragmentation (Lijphart, Pintor et al. 1986). In Afghanistan it produced wildly erratic results. Small vote differentials could separate winners from losers, and "some candidates won with tens of thousands of votes while others scraped through with as low as 1,500" (Coburn and Larson 2014: 58). This produced resentment among candidates who were close to being elected, and the electorate saw an extraordinarily high rate of "wasted" votes for candidates who were ultimately not elected (Coburn and Larson 2014: 58–59).

The 2010 Parliamentary Elections

These trends continued during the 2010 elections. The growing threat of Taliban violence impacted voter turnout, and parliamentary candidates were targeted for assassination. Through an expansive and legally dubious decree, the president unilaterally reformed the election law (Coburn and Larson 2014: 197–200). Karzai actively undermined the Electoral Complaints Commission, which "had invalidated over 1.3 million fraudulent votes cast in the 2009 presidential poll," by replacing most of its members with presidential appointees (Maley 2011: 86). Unsurprisingly, fraud was rampant. Significant delays in certifying the winners lead to speculation that the vote was rigged. The electoral volatility continued as over half the sitting Wolsei Jirga members lost their seats (Ginsburg and Huq 2014: 117). Once again, elections resulted in a highly fragmented Wolsei Jirga, feckless political parties, popular dissatisfaction, and no discernible progress toward the rule of law.

The Legal System

After the Taliban's collapse in late 2001, the state legal system was in shambles. By 2014 there were some advances, but progress was underwhelming given the resources invested. A legal structure existed for the judiciary. The Organization of the Court Law centralized power in the Supreme Court (Islamic Republic of Afghanistan 2005). The law envisioned appellate courts in each province and hundreds of district-level courts. Hundreds of courts became operational, and

> significant work has been done on legislation; several hundred judges, prosecutors, and prison wardens, and thousands of police personnel have been trained; some justice institutions have been refurbished; and several new ones have been built from scratch. Progress has also been made with regard to building administrative capacity within the existing justice institutions and the publication and distribution of a large body of law to legal professionals. Progress in rebuilding Afghanistan's state justice system has included the establishment of the Independent Bar Association of Afghanistan, legal aid departments in Kabul and in three provinces, the Independent National Legal Training Centre (INLTC) in Kabul, and a committee for the simplification of judicial bureaucracy. (Wardak 2011: 1308)

While state justice competed with powerful non-state judicial authorities, it was not committed to promoting the rule of law. The state legal system was immensely corrupt, ineffective, expensive, and arbitrary. Court decisions were often unenforceable. The legal system mirrored the larger state apparatus in that it was most powerful in Kabul and some other cities while largely nonexistent in more rural areas. Unsurprisingly, most people sought to avoid it. Disputants sometimes even threatened adjudication in state courts to facilitate non-state dispute resolution (Local NGO Professional 2014; NGO

Legal Specialist 2014). The state judiciary's authority waned even as its personnel and institutions proliferated.

In practice, the judiciary was not independent. It consistently upheld executive authority, including in conflicts over the scope of executive power, election administration, and the removal of presidential appointees (Hamidi and Jayakody 2015). The Supreme Court bolstered executive power rather than the rule of law. As one judge explained, even if a judge personally wanted to follow the law, the incentives were perverse. Judicial posts were routinely bought and sold.[14] Judges were not particularly well-paid, and they often attempted to recoup the costs of buying their post (Former Afghan Judge 2014). Judges were rarely stationed in their home area, which meant they frequently did not speak the local language or understand the local context. Adjudication was often corrupt and time-consuming, and enforcement capacity was lacking in most areas. Prosecutors routinely refused to prosecute well-connected individuals but used the threat of prosecution to extract rents (Afghan International Rule of Law Program Staffer 2014; Chayes 2015). Likewise, the subsidiary institutions designed to support the judicial branch, most notably the legal education system, consistently left graduates unprepared to practice law (Swenson and Sugerman 2011).

Under conventional judicial review theory, the judicial branch determines what is legally permissible (Kramer 2001; Merryman and Pérez-Perdomo 2007). In Afghanistan the Supreme Court and state judicial courts bolstered presidential power in a paradigmatic example of rule by law. Presidential appointments did not necessarily follow the law even for the Supreme Court. Karzai made the first judicial appointments unilaterally under the authority granted by the Bonn Agreement. These appointments were subsequently entrenched by Article 127 of the 2004 constitution. The first chief justice was the conservative Islamist Fazi Hadi Shinwari.[15] After parliament rejected Shinwari's renomination in 2006, Abdul Salam Azimi assumed the post. The Wolsei Jirga initially approved the appointment. After Azimi's term expired in 2010, Karzai retained him, giving him the extraconstitutional title of "acting chief justice," along with two other judges, "through a legally dubious decree" (International Crisis Group 2012: 14).[16] The appointment

[14] The selling of state posts was by no means limited to the judiciary (Goodhand 2008).

[15] Shinwari's 2004 Supreme Court appointment was also constitutionally questionable. The then-controlling 1964 constitution stipulated that all judges had be under sixty years of age (Kingdom of Afghanistan 1964). At the time of his appointment, however, Shinwari was eighty years old.

[16] Once Karzai left office, Azimi resigned in October 2014.

power coupled with the retirement power and the other executive powers reinforced executive authority over the judiciary.

State Attempts to Regulate Non-State Justice

Non-state justice mechanisms still resolved most disputes. Developing a constructive relationship with non-state legal authorities, however, was not a state priority. On one hand, a minimalist approach made sense. After all, state capacity to enforce court judgment or regulate non-state actors was limited, particularly outside urban areas. Poorly designed regulations can do far more harm than good. Unfortunately, inaction created a legal gray zone. Contested issues resolved by non-state forums, such as land or property, remained formally unsettled as their agreements had no status in state courts and thus can be subject to further litigation.

From 2006 on interest in the non-state justice sector grew considerably. In 2009 a state policy was crafted that envisioned simultaneous harmonization and incorporation strategies. All major state justice sector institutions, including the MOJ, the Supreme Court, the Human Rights Commission, and the Ministry of Women's Affairs supported the 2009 informal dispute resolution policy. It also enjoyed support from key international actors and NGOs. At no point, however, did state officials seriously engage major non-state justice actors.

The policy had limited scope and sought to encompass diverse goals that were often in tension with each other. The policy on informal dispute resolution covered only shuras and jirgas (Ministry of Justice 2009). While recognizing shuras and jirgas as "critical components to the establishment of long-term peace, security, and rule of law," it stressed that in order to be binding and valid, non-state legal decisions had to be consistent with sharia, the constitution, other Afghan laws, and international human-rights standards (Ministry of Justice 2009). The policy called for the elimination of all practices violating human rights, mandated gender equality, and made participation in non-state justice processes completely voluntary. Jurisdiction was sharply curtailed, such that it was restricted to civil cases that did not concern the government.

Whatever policy consensus existed was shallow and came at the expense of feasibility and recognition of how non-state justice actually worked. The policy assumed the state could simply impose its mandates and that non-state actors would proactively and uncritically support a state judicial system that was widely known to be predatory, corrupt, and ineffective. Moreover, no clear vision existed to address the de facto authority of warlords in many parts of the country (Malejacq 2019).

The September 2010 Draft Law on Dispute Resolution Shuras and Jirgas sought to operationalize the 2009 policy. It envisioned non-state justice actors' becoming part of the state system. Consequently, it heavily regulated their jurisdiction, operations, and decision-making as well as their relationship to state courts (Ministry of Justice 2010). As such, the draft law included harmonization and bridging elements alongside incorporation. The law staunchly asserted the state's authority to control and regulate all aspects of non-state dispute resolution. In addition, it stipulated that jirga participants "and parties of dispute shall be duty bound to observe provisions of this law" or face criminal charges, even though jirgas were empowered only to hear civil disputes and petty juvenile crimes on referral (Ministry of Justice 2010). Jirgas and shuras could not "make decisions that violate [the] human rights of parties in dispute, especially of women and children," and all judgments would have been subject to appeal to state courts (Ministry of Justice 2010). The law ultimately was not passed owing to fierce opposition from the Ministry of Women's Affairs and the Human Rights Commission, which feared that it would lend credibility to "traditional" dispute resolution, something they view as antithetical to human-rights standards and women's rights (International Rule of Law Professional 2015).

State officials were open to engagement with the non-state justice sectors, but only on their own terms. The central government's internal divisions ultimately prevented the law from being promulgated. Indeed, no law was ever passed prior to the collapse of the regime in August 2021. Even if a law had been enacted, it would have been unlikely to have meaningfully altered how dispute resolution operated locally, given the weakness of state authority in many areas.

Judicial State-Building Competitors and Combatants

Poor governance and the lack of a legitimate legal order are necessary though not sufficient conditions for anti-state insurgencies to flourish (Fearon and Laitin 2003; Hironaka 2009). During the relatively peaceful interlude after the Taliban regime's collapse, little was done to promote the development of a legitimate state bound by the rule of law. Instability spread quickly. By 2006 the Taliban resurgence "had developed into a full-blown insurgency," and state authority has steadily eroded since then (Jones 2008: 7), a situation that reflects the challenges of establishing the military and civilian authority structures necessary to maintain order after conflict as well as the regime's policy choices. A neo-Taliban insurgency may have been inevitable, but many fighters were initially willing to join the new regime (Johnson 2011: 259).

The Karzai regime could not defeat its competitors militarily and did not represent a cause worth fighting for. Nowhere is this failure more pronounced than in the justice sector, where the state failed to promote or even provide a just legal order. The absence of a legitimate, fair, and effective state legal system had major consequences. Many Afghans continued to use tribal dispute resolution mechanisms, particularly in Pashtun tribal areas (Wardak and Braithwaite 2013). Cumulatively, non-state justice mechanisms settled an estimated 80 to 90 percent of disputes (Barfield, Nojumi et al. 2006: 9).

Warlords

The Bonn Agreement and the political-party law envisioned the disarmament of the warlords. Reality, however, was starkly different. The Karzai regime and international forces relied on "pro-government or more accurately 'anti-Taliban' warlords to maintain order" both locally and regionally (Jones 2010: 130). Halfhearted efforts to break up the military underpinnings of the warlords' power failed. This is hardly surprising. As Sedra notes, "A DDR [disarmament, demobilization, and reintegration] program has never been successfully implemented without the support of a state or external security force to assist its work" (Sedra 2003: 8). A network of warlords quickly became entrenched, one that contested state authority but focused on their spheres of influence rather than trying to seize control of the state itself (Peceny and Bosin 2011).[17]

The Taliban

Despite losing power in 2001, the reconstituted Taliban quickly established itself as the state's fiercest judicial rival. Taliban justice became a centerpiece of a full-blown insurgency by 2006 that sought to displace the state itself

[17] Autonomy should not be confused with isolation. Warlords such as Atta Mohammad Noor in Balkh, Gul Agha Sherzai of Nangarhar, and Ismail Khan of Herat frequently engaged the state and even held state posts when it suited their interests (Mukhopadhyay 2014).

(Giustozzi and Baczko 2014). The Taliban established a lean but sophisticated network of parallel governance structures. Once territory started falling under the Taliban's control in 2003, provincial governors were almost immediately established, and there were an estimated "33 provincial governors and about 180 district governors" by 2010 (Giustozzi 2012b: 72).

Effective legal order constituted the core of the Taliban's political program, underpinned their claim to be Afghanistan's legitimate rulers, and highlighted the state justice system's failures. The Taliban emphasis on legalism was also evident in its published codes of conduct (Layha), which started in 2006. These edicts sought to regulate appropriate behavior for Taliban fighters in both military and political matters. While the code is rarely followed to the letter, Clark contends that it is not purely "propaganda: it is a rule book that is also aspirational" (Clark 2011: 2). The Taliban even established military courts to try to implement the code (Giustozzi 2014). Taliban insurgents actively contended with the state legal system, especially outside the capital. The Taliban operated "a parallel legal system that is acknowledged by local communities as being legitimate, fair, free of bribery, swift, and enduring" (Johnson 2013: 9). Unlike in the state court system, decisions were widely enforced (Ledwidge 2009; Kilcullen 2011). Moreover, their justice system was "easily one of the most popular and respected elements of the Taliban insurgency by local communities, especially in southern Afghanistan" (Johnson 2013:9). Taliban justice sought to provide what the state justice system did not: predictable, effective, legitimate, and accessible dispute resolution.

Mitigating corruption was a top priority, and to this end an ombudsman system was included where citizens could make complaints. As Peters observed, "The Taliban's willingness to punish their own" is a major reason the general population saw them as "more fair—even if strict and ruthless—than the notoriously corrupt Afghan government" (Peters 2011: 108). While the Taliban's justice system had major shortcomings and its human-rights record was appalling, the Taliban possessed a comprehensive legal strategy with a clear long-term vision rooted in local values and beliefs.

The Taliban explicitly rejected the constitutional foundation of the post-Bonn Afghan state (International Crisis Group 2020). Afghan and international forces actively sought to disturb and undermine Taliban justice system that posed a profound challenge to state authority (Coburn 2013). Even North Atlantic Treaty Organization (NATO) officials begrudgingly recognized the Taliban judiciary's relative effectiveness:

> While Afghans disagree with the harsh punishments of the Taliban, they often find this "extreme justice" their sole recourse for injustice. The redress of grievances is one of the few areas where the insurgency continues to compete with legitimate governance. (NATO 2011)

The Taliban sought "to discredit and undermine" state claims to legal authority (Johnson 2011: 282). They tried to disrupt the operations of state courts and targeted judges and other state officials for assassination; they also denied the courts' legal authority. Populations in Taliban-controlled areas were banned from utilizing state courts and faced punishment for engaging state authorities (Johnson and Mason 2007; Giustozzi and Baczko 2014). Thus, the relationship between the state and Taliban justice systems was a paradigmatic example of combative legal pluralism.

The Taliban were equally committed to destroying their warlord competitors (Mac Ginty 2010). This process was also combative, though there was perhaps less focus on quality of services and more emphasis on battlefield victories. Nevertheless, the political and ideological portions of the Taliban's justice strategy remained important in areas it controlled as well as those it aspired to control (Rashid 2008: 362–363).

Unlike the state legal system, the Taliban justice system capitalized on the wellspring of legitimacy that it used so effectively in the mid-1990s: the provision of order backed by religion and tribal identity. It emphasized its religious piety, in contrast to the Kabul-based regime. The Taliban judiciary's claim to adjudicate based on sharia law "strengthens their legitimacy in a deeply religious population, particularly when the codes of law used by the state are little known, misunderstood, and sometimes resented" (Giustozzi and Baczko 2014: 219). Religion also underpinned efforts to undermine the state's legal authority and Taliban attempts to mobilize popular opposition to the state (Jones 2010: 230–237).

The Taliban stressed their respect for local values, traditions, and realities. At the same time, they were pragmatic. Shuras and jirgas remained a key part of the legal landscape. Although focused on religion, the Taliban are also deeply imbued with Pashtun tribal values and often work constructively with tribal leaders:

> The Taliban judges often co-opt the elders into their trials, allowing them to represent the parties and to negotiate a settlement. In practice, the system adopted by the Taliban has made the functioning of customary justice

> increasingly dependent on the Taliban's support, without which communities would often not be able to enforce decisions. (Giustozzi and Baczko 2014: 212)

Due to this dependence and the strength of the insurgency, many of these forums faced growing pressure to operate in a manner deemed appropriate by the Taliban (Jackson and Weigand 2020). As an armed insurgent group, the Taliban employed a strategy that coupled the promise of legal order with the threat of force. Johnson and Mason offer a vivid description of Taliban threats:

> By night, Taliban mullahs travel in the rural areas, speaking to village elders. They are fond of saying, "The Americans have the wristwatches, but we have the time." The simple message they deliver in person or by "night letter" is one of intimidation: "The Americans may stay for five years, they may stay for ten, but eventually they will leave, and when they do, we will come back to this village and kill every family that has collaborated with the Americans or the Karzai government." Such a message is devastatingly effective in these areas, where transgenerational feuds and revenge are a fabric of the society. (Johnson and Mason 2007: 87)

The Taliban's success at judicial state-building dispels the notion that establishing legal order was impossible in post-2001 Afghanistan. Despite extensive efforts to suppress it, the Taliban legal order continued to gain ground and popularity (Weigand 2017). The Taliban offered a compelling strategic vision for many Afghans. Its legal order blended overt coercion with a serious attempt to ground proceedings in the cultural and religious basis of legitimacy along with relatively consistent judicial procedure and application of the law (Provost 2021: 118–140). Coercion can be deeply unpleasant and provoke resentment. At the same time, enforcement of judgments constitutes the core of an effective legal system. This is manifested in the belief that the law will be applied to all. All effective legal systems, at a minimum, require authority to mediate competing interests and some capacity to enforce their rules on those who fail to comply (Raz 2009).

Coercion alone cannot sustain a legal or political order in the long run, because stability requires a degree of cooperation (Beetham 2013). Societal respect for the law largely hinges on the broad social belief that it is basically fair and legitimate (Tyler 2006). The Taliban's legal order followed a template that has already proven effective historically: a guarantee of order backed by respect for religious and cultural beliefs. The Taliban's legal order is further bolstered by at least a limited willingness to police itself and be (somewhat) bound by their own law (Johnson and DuPee 2012).

The Taliban's justice system, however, should not be idealized. It wholeheartedly rejected the human-rights and cultural requirements embodied in a thicker version of the rule of law and explictly discminated against women. Whatever virtues the Taliban legal system possessed were primarily in comparison to the state legal system. The Afghan people quite correctly saw the Afghan judicial system as an "extractive institution" that collected rents from society at large and redistributed them to a narrow elite (Acemoglu and Robinson 2012: 81).

Through a mixture of force and provision of a vital public good, legal order, the earlier Taliban regime transformed an environment defined by competitive legal pluralism into one of cooperative legal pluralism, wherein local authorities willingly worked with them or, at a minimum, did not actively oppose them. In contrast, the state system in most areas lacked coercion and legitimacy as well as uniformity and accountability. Unlike in the period from 1996 to 2001, the Taliban during this period did not establish a monopoly on legitimate violence, but they did help lay the foundation for their ultimate success. Taliban justice faced serious pressure from state and international efforts to repress it. It not only survived but thrived. The Taliban insurgency succeeded owing in large part to the Afghan state's abysmal performance. Now that the Taliban has returned to power, it will be telling to see whether the Taliban justice system has internalized the lessons of the regime pre-2001. In the mid-1990s the Taliban rose to power promising fair, accessible, and legitimate dispute resolution, but popular support waned as Taliban rule became ever harsher and more unaccountable. Discontent could certainly swell once again.

Conclusion

Afghanistan is certainly a difficult place to govern. But it is not predestined to lawlessness. Throughout Afghanistan's history, enduring pillars of legitimacy—cultural values, Islam, and the provision of public goods—have guided effective legal orders and governance more generally. Despite the upheaval and violence the country has endured since 1978, these building blocks of an effective, legitimate legal order remain in place. The Karzai regime choose to eschew these long-standing pillars of legitimacy. While promising a new beginning, the state became thoroughly corrupt, predatory, and fundamentally illegitimate. It also did not seek to work constructively with non-state judicial authorities rooted in tribal structures.

After 2014 state-backed law and governance continued to deteriorate. Afghanistan underwent a major transition following the election late that year of President Ashraf Ghani. Unfortunately, Afghanistan's first peaceful transfer of power was compromised by massive election fraud conducted on Ghani's behalf by the outgoing President Karzai. Ghani pledged extensive political, legal, and electoral reform but accomplished little. The Taliban's territorial control continued to grow, and by the end of 2014 the Taliban controlled roughly 50 percent of Afghanistan (Almukhtar and Yourish 2015). Taliban justice continued to proliferate in the run-up to the movement's ultimate victory (Jackson and Weigand 2020). The regimes led by President Karzai and his successor, President Ghani, deserve most of the responsibility for Afghanistan's decline since 2001 and its fall in 2021. As the next chapter highlights, however, the international community not only helped install and consolidate a regime that actively opposed the development of democracy or the rule of law; it also enabled many of its abuses.

7
International Subsidization of a Rentier State in Afghanistan

Introduction

Chapter 6 argued that judicial state-building failures resulted largely from the decisions of key state policy makers. The Karzai administration established a highly centralized, personalized regime antithetical to the rule of law. However, the state did not act alone. Afghanistan received an extraordinarily high level of international support from 2002 to 2014. In fact, the Afghan state depended on international support to undertake even its most basic of functions (U.S. Government Accountability Office 2013: 25–26). In the justice sector, roughly 90 percent of funding came from foreign sources (Suhrke and Borchgrevink 2009b; Leeth, Hoverter et al. 2012: 4).

Choices about how and where to provide international assistance can have major consequences (Swenson and Kniess 2021). In Timor-Leste it made a significant positive contribution in a helping transform a competitive legal-pluralist environment into a cooperative one. Aid, however, can be ineffective or even counterproductive to promoting a democratic state bound by the rule of law. This chapter argues that, on balance, donor assistance did little to promote the rule of law or support a democratic state. Nor did it help to build a more cooperative relationship between state and non-state justice in Afghanistan. Progress toward the rule of law proved elusive despite billions in aid, the international community's seemingly immense leverage given the state's donor dependence, and, after 2009, an increased willingness to engage non-state judicial actors. At best, assistance produced mixed results. It supported key legal institutions and improved human-resource capacity. At the same time, aid was often ad hoc, ineffective, predicated on widely optimistic assumptions and, at worst, helped enable predatory state practices. Shortly before the Islamic Republic of Afghanistan collapsed entirely, the justice system was still widely seen as the most corrupt part of an extraordinarily corrupt state (McDevitt and Adib 2021). In contrast, the Taliban's

Contending Orders. Geoffrey Swenson, Oxford University Press. © Oxford University Press 2022.
DOI: 10.1093/oso/9780197530429.003.0007

rival justice system continued to thrive. This chapter examines the key question: Why, despite, major international investment in judicial state-building, did the quality of the justice sector not improve?

Summary of Arguments

In Afghanistan, high-level state officials, particularly President Karzai, both mediated international assistance and determined the overarching trajectory of judicial state-building. In the early years, however, the international community exercised tremendous influence over the nascent Afghan state's institutional structure. International backing secured the top state post for Karzai and encouraged the concentration of power in the executive. Foreign actors retained significant ability to influence government officials because the state was highly dependent on their aid. Yet, international judicial state-building activities did little good and were occasionally even counterproductive to advancing the rule of law and fostering constructive engagement with non-state actors. International actors never truly sought to engage with non-state justice as a means to promote a more just legal order; rather, they saw non-state actors as mere cogs in a deeply flawed counterinsurgency approach.

Conventional Understandings of Judicial State-Building in Afghanistan Are Insufficient

There is widespread agreement that despite massive expenditure, international assistance did not promote the rule of law or even mitigate the corrupt and predatory nature of the new regime. Yet, many scholars contend the result was basically inevitable (Ottaway and Lieven 2002; Suhrke 2011). Acemoglu and Robison, for instance, posit that the failure of assistance to Afghanistan "should not have been a surprise, especially given the failure of foreign aid to poor countries and failed states over the past five decades" (Acemoglu and Robinson 2012: 451). This chapter argues that, despite serious challenges, as with the domestic efforts discussed in Chapter 6, international judicial state-building efforts were not predestined to fail. There were clear policy and program choices that had profound consequences. This helps provide a fuller picture by looking at international judicial state-building holistically and at a programmatic level. International aid rested on faulty assumptions, made serious miscalculations, and could have been done better in relation both to how the state justice sector was supported and how the non-state justice sector was engaged. This examination reveals that while success can never be guaranteed, it can be undertaken in a way that substantially improves the likelihood of success.

Evaluating International Judicial State-Building Assistance

As with the earlier evaluation of international judicial state-building assistance in Timor-Leste, success is assessed based on whether international aid *enhanced the prospects for developing and consolidating the rule of law.* International rule-of-law assistance is analyzed through a process-tracing approach of major initiatives. Chapter 6 illustrated how the Karzai administration actively undermined domestic efforts to promote state-building based on the rule of law. The regime effectively undermined domestic checks and balances and the separation of powers. Perhaps most troublingly, the regime compromised the electoral system through which the population exercises sovereignty and also limited the development of representative political parties and an independent, professionalized state judicial system. The state also never meaningfully engaged with the crucial religious and tribal non-state justice actors who have long underpinned legitimate legal and political order in Afghanistan.

From the start the international community viewed security as its top priority, which relegated the rule of law to a secondary concern (Wolfowitz 2002). Lakhdar Brahimi, the chief U.N. representative at the Bonn Conference and head of the U.N. mission in Afghanistan from March 2002 to January 2004, explicitly stressed that "security is more important than justice" (Brahimi, cited in Larsen 2010: 21). International support for establishing the rule of law was consistently trumped by its commitment to bolstering Karzai's regime under the guise of ensuring security.

Paradoxically, despite the justice sector's extreme dependence on aid, international actors' willingness to promote the rule of law was quite limited. In theory, aid dependence should increase the state's reliance on the continued flow of aid and therefore allow major donors meaningful influence over state behavior (Boyce 2002; Knack 2004). Most donors want to avoid being seen as dictating state behavior and emphasize "local ownership" (see, e.g., USAID 2005; UNDP 2009a; U.S. Mission Afghanistan 2010). The Karzai government believed that the United States and the broader international community were not prepared to allow the regime to collapse owing to their fear that Taliban would return. Over time, the state's endemic weakness gave the regime greater ability to operate with independence from its international backers. Indeed, this dynamic continued until the Biden administration decided in 2021 that the United States was not willing to prop up the Afghan government indefinitely.

Rule-of-law program implementers also faced structural constraints. Donors assessed implementers based on their ability to execute stipulated programming deliverables, which almost always involved working with state institutions. This dynamic demanded state consent regardless of whether officials were actually committed to programmatic principles. Programs that failed to deliver were viewed negatively by donors, and consequently future contracts could be endangered (Cooley and Ron 2002). Program implementers had strong incentives to collaborate with deeply corrupt state institutions, provided their leaders were willing to at least pay lip service to rule-of-law ideals. These problems were further compounded by the steadily deteriorating security situation, which limited the areas and types of work that could be done at an acceptable level of risk.

Chapter Overview

This chapter is organized thematically. The first section looks at the major international judicial state-building efforts and whether each major program enhanced the prospects for developing and consolidating the rule of law. Given the vast number of initiatives, it is impossible to examine each program in detail. The State Department's inspector general explained the inherent challenges in disentangling even just U.S. assistance:

> ROL funding is difficult to identify and to quantify. Funding for the ROL program in Afghanistan is split among several U.S. government agencies. There is no one place where all funds spent specifically on ROL can be identified. ROL program funding is often multiyear and is combined with other programs such as police training and correction facilities, which often make identification of specific costs difficult. ROL programs are also funded by the United Nations, other bilateral donors, and a variety of NGOs. The result is that there is currently no way to readily identify ROL funding and subsequently to identify duplicate programs, overlapping programs, or programs conflicting with each other. (U.S. State Department Office of the Inspector General 2008: 23)[1]

[1] Strategy, implementation, and coordination problems remained endemic despite repeated claims that they were being addressed (SIGAR 2015b).

Nevertheless, the key programs can still be identified and evaluated. The first section surveys the work of the most important international actors. It begins by examining Italy's work as the "lead" donor nation for the justice sector. This is followed by an evaluation of major U.S.-funded efforts. This section concludes with a discussion of U.N. and UNDP initiatives. The second section holistically examines the strategy and trajectory of externally funded judicial state-building projects before assessing their larger impact upon institutions and the broader justice sector more generally.

International Actors and Initiatives

A wide variety of countries and international organizations have supported judicial state-building in Afghanistan since 2001.[2] Unlike in Timor-Leste, where a significant amount of assistance was rooted in cultural and linguistic affinity, nearly all of the international assistance to Afghanistan has been conventional assistance seeking to develop a modern, high-capacity justice sector. The main international actors were the United States, the United Nations, the UNDP, and, at least initially, Italy, which was the lead nation charged with "providing financial assistance, coordinating external assistance and overseeing reconstruction efforts in the rule-of-law sphere" (Tondini 2007: 336).[3]

Italy

Italy was formally the lead nation for the justice sector from 2002 to 2006.[4] It had the largest role in providing financial aid, overseeing external assistance, and coordinating projects.[5] Italy focused on reforming and expanding legislation and the state legal system. Italian aid sought to reestablish the state justice sector in a manner similar to what existed prior to the communist takeover, as well as to promote human-rights norms. Support focused on "(*i*)

[2] For an overview of the Afghan institutions and initiatives that received international support, see Afghanistan Research and Evaluation Unit (2015).

[3] Germany also funded rule-of-law work in Afghanistan starting in 2003 (GTZ 2014).

[4] The other lead nations were: the United States, armed forces; Germany, police; Japan, disarmament; and the United Kingdom, counter-narcotics.

[5] The lead-nation approach was formally eliminated after the 2006 London Conference on Afghanistan.

legislative reform, (*ii*) training and capacity building, and (*iii*) rehabilitation of infrastructure" (Tondini 2010: 48). Initially, Italy implemented this work through state agencies, most notably the Italian Justice Project Office and the International Development Law Organization (IDLO).

The initial post-intervention years after the fall of the Taliban were the period of "maximum possibility" (Dobbins, Jones et al. 2007: 15). Yet, Italy "took a reactive, almost passive position" while it supposedly waited for guidance from the relevant Afghan authorities (Suhrke 2011: 188). When assistance commenced in earnest in February 2003, it was very technocratic. Support focused on locating legislation lost during decades of conflict and drafting new legislation, including the Juvenile Code, the Law on Prisons and Detention Centers, and particularly the Interim Criminal Procedure Code (ICPC).

Italy determined that war-ravaged Afghanistan's most pressing judicial need was a new ICPC, despite an extant code from Daoud's era. The end result was a "brief version of the Italian civil procedure code" (Suhrke 2011: 189). Italian officials stressed the ICPC was "a simplified text designed to make the work of the criminal police and judges easier and compliant with international human rights" (cited in Ahmed 2005: 101). Under intense political pressure, Karzai promulgated the ICPC in early 2004. Italy apparently even threatened to withhold development assistance worth millions if the ICPC was not approved (Hartmann and Klonowiecka-Milart 2011: 276). The law, however, generated domestic immense opposition from the MOJ and the broader legal community, who believed it was a poor fit for the Afghanistan legal environment. The Italian-dominated drafting process for the ICPC was deeply flawed and showed little interest in the views of Afghan stakeholders. Substantively, it more or less replicated Italy's criminal procedure code (Hartmann and Klonowiecka-Milart 2011). Afghans saw the ICPC as an outside intervention that failed to constructively engage with the existing legal framework or have a credible implementation plan. The law stipulates that "any existing laws and decrees contrary to the provisions of this code are abrogated" without identifying those laws or if the ICPC repeals or supplements the preexisting code (Islamic Republic of Afghanistan 2004b). The law was a decided failure (Röder 2007: 309), and a new law was eventually drafted to replace it.

Italy's performance has been rightly criticized for its initial inaction followed by an ill-conceived, heavy-handed approach. As noted by a major RAND study, "Italy simply lacked the expertise, resources, interest, and influence needed to succeed in such an undertaking" (Dobbins, Jones et al. 2007: 1).[6] Italian assistance did little to bolster the rule of law and squandered a crucial window of opportunity. Moreover, it failed to constructively engage with the pillars of judicial legitimacy, most notably tribal and religious authority, when the timing was most auspicious for collaboration.

United States

While Italy was the lead nation for the justice sector, the United States was a major player. After Italian assistance was scaled back, the United States more than filled the void. Since 2003 the United States has funded a slew of major initiatives through various government agencies, most notably USAID, the State Department, and later the Department of Defense (DOD).[7] A major 2015 study found that the "DOD, DOJ, State, and USAID have spent more than $1 billion on at least 66 programs since 2003 to develop the rule of law in Afghanistan" (SIGAR 2015b: 1). As with the overarching U.S. state-building effort in Afghanistan, each agency involved pursued its own programmatic priorities, which were not necessarily coordinated or even broadly consistent (Keane and Diesen 2015).

Key USAID Initiatives

USAID–Afghanistan Rule of Law Project (AROLP), 2004–2009

USAID became the first major provider of U.S. rule-of-law assistance with the $44 million Afghanistan Rule-of Law (AROLP) program.[8] Checchi Consulting operated AROLP from October 2004 to July 2009. The initiative focused on (1) "strengthening court systems and the education of legal personnel," (2) reforming the legislature, (3) improving access to justice, and (4) eventually engaging with the informal sector (Checchi and Company Consulting 2006: 3). Reforming the commercial courts as well as promoting human rights and women's rights under Islam were also priorities. Capacity-building formed the cornerstone of programming. The program instructed

[6] Other assessments of Italy's performance are also negative, see International Crisis Group (2008: 6); Sherman (2008: 323); Hartmann and Klonowiecka-Milart (2011); and Suhrke (2011: 186–192).

[7] All figures for comprehensive program costs were taken from SIGAR (2015b).

[8] See SIGAR (2015b) for an overview of all 14 USAID programs related to the Afghanistan's justice sector. In total, USAID invested over $13.3 billion in Afghanistan reconstruction efforts during this period, of which rule-of-law efforts formed only a small subset (SIGAR 2014c).

more than one thousand judges as well as approximately sixty law professors and one hundred university administrators (Checchi and Company Consulting 2009b: 3). AROLP produced legal materials and sought to improve the courts' administrative capacities. As with the vast majority of rule-of-law assistance, it was highly technocratic and state-centric.

In its later years, as state-focused initiatives were already beginning to fall short of expectations, AROLP displayed some interest in the non-state justice sector. Programmers believed that "the linkage between the formal and informal justice sectors [was] essential to improving the rule of law in Afghanistan" (Checchi and Company Consulting 2009b: 33). AROLP sponsored research on non-state justice and tried to increase citizens' awareness of the state justice system, along with "their legal rights and responsibilities under the Constitution of Afghanistan" (Checchi and Company Consulting 2009b: 33). The program also advocated for "a national policy on state relations with informal justice mechanisms" (Checchi and Company Consulting 2009b: 34).

While AROLP worked with various justice-sector institutions, the Supreme Court was its most important partner. The Supreme Court's consent was essential for most programming, but, as discussed in Chapter 6, the judiciary was neither independent or committed to the rule of law.[9] As a result, Supreme Court judges had little incentive or ability to foster the rule of law. The Supreme Court consistently upheld executive authority, however dubious the exercise of that power, including in conflicts over the scope of executive power, election administration, and the removal of presidential appointees (Hamidi and Jayakody 2015). AROLP did little to address structural problems in the Afghan justice sector or constructively engage the non-state sector. This dynamic is evident based on the program's chosen evaluation criteria: (1) creation of a "national policy on the informal justice sector" and (2) usage of state courts as reflected in survey results (Checchi and Company Consulting 2009a: 25). A non-state justice policy was developed by domestic and international actors, marking the first official state recognition of non-state justice authorities. Despite formal agreement at the ministerial level, the policy did not reflect a real consensus because it relied on ambiguous language to paper over major differences and did not involve any serious engagement with non-state judicial actors.

[9] AROLP also worked with lower courts, but under the direction of the Supreme Court.

The second evaluation criterion was equally problematic. First, projects were predicated on the simplistic and inaccurate belief common to rule-of-law programs that informing people about the state courts and their rights will improve the rule of law (Carothers 2003). Second, the program sought to channel people to state courts regardless of the consequences. Courts were initially underresourced, both financially and in terms of human resources. They were also slow, expensive, and possessed limited enforcement capacity. Worse, they were often corrupt, predatory, and rent-seeking (De Lauri 2013; Singh 2015). It made little sense to direct cases into the state justice system regardless of the quality and capacity of its courts. As the first major USAID rule-of-law program, AROLP did next to nothing to advance the rule of law.

USAID Rule of Law Successor Programs (2010–2014)

By 2009 promoting a just, viable legal order was increasingly seen as a potent weapon against an increasingly powerful Taliban insurgency that drew strength from the Afghan government's immense corruption. USAID funded two related rule-of-law programs that rejected the artificial opposition between security and a just and effective legal system.[10] Rule of Law Stabilization–Formal Component (RLS-Formal) sought to "develop[] a justice system that is both effective and enjoys wide respect among Afghan citizens is critical to stabilizing democracy and bringing peace to the country" (Tetra Tech DPK 2012: 1). Rule of Law Stabilization–Informal Component (RLS-Informal) shared a similar premise, as it aimed "to 'promote and support the informal justice system in key post-conflict areas' as a way of improving stabilization" (Scope of Work, Annex A in Dunn, Chisholm et al. 2011: 35).

RLS-Formal

The international development contractor Tetra Tech DPK was chosen to implement RLS-Formal from May 2010 to September 2014, at a cost of more than $47.5 million.[11] The program's goals were strikingly similar to those of AROLP. RLS-Formal emphasized "capacity building" of judicial

[10] Following a controversy over initially awarding the entire successor rule-of-law program as a no-bid contract to Checchi, USAID split its rule-of-law support programs into two major strands: formal and informal, with two separate implementers (Coburn 2013: 41).

[11] While both the formal and informal programs ran from 2010 to 2014, there were three distinct phases, as USAID did not immediately approve a four-year program. Rather, there were two one-year programs and a two-year program. The objectives remained the same, but it did negatively impact program planning and continuity.

actors and administrators, as well as improving legal education and raising "public legal capacity and awareness" (Leeth, Hoverter et al. 2012: 1; Tetra Tech DPK 2014).[12] RLS-Formal supported extensive training and technical assistance programs, along with the development of educational materials and strategic plans. There were also public information campaigns and attempts to strengthen the public outreach capacity of the Supreme Court and the MOJ. Most participants viewed the training positively, but RLS-Formal faced chronic difficulties (Leeth, Hoverter et al. 2012). The program emphasized "[improving] the public image and use of the formal judiciary," but improving public perception risked enabling rent-seeking (Leeth, Hoverter et al. 2012: 6). RLS-Formal ignored structural issues with the underlying political economy of the Afghan justice system, including the lack of judicial independence and endemic corruption. As before, public service announcements encouraging more people to use a highly corrupt justice system did little to improve the rule of law and threatened to further undermine Afghans' faith in the judicial system.

Programming, again, depended on consent from the institutional actors responsible for many of these problems. These actors had little incentive to facilitate changes that threatened the larger political and legal order. Tetra Tech's final report is admirably forthright in this regard. After noting major challenges with corruption and "weak political support for rule of law reforms," the report explains:

> The most difficult barriers to overcome during project implementation, and the barriers that threaten the sustainability of project initiatives, are a lack of willingness among counterpart institutions to support and adopt reforms, and a failure to allocate sufficient funds to maintain quality trainings for current and future justice-sector personnel. Leadership at counterpart institutions continues to demonstrate a lack of commitment to justice sector reforms by delaying the approval of tools and technologies that will increase the efficiency and transparency of courts. (Tetra Tech 2014: 3)

The Supreme Court was particularly problematic, because it received and oversaw external aid but lacked a commitment to the rule of law. As Tetra Tech observed, "When the Supreme Court was asked to approve new

[12] The program also undertook activities to promote gender equality and other social goals.

initiatives or reforms, the institution tended to make approval contingent on future events that never happened" (Tetra Tech 2014: 3). It determined that the Supreme Court's attitude reflected a clear "reluctance to embrace new processes and procedures that increase the efficiency, transparency, accountability, and fairness in the justice sector" (Tetra Tech 2014: 3).[13] Thus, RLS-Formal initiatives had few prospects for sustainability or fostering meaningful institutional change.

Although Tetra Tech implemented the program, USAID deserves the most strident criticism. It designed the program and continually displayed almost willful blindness to political reality. As the SIGAR noted, "USAID nearly doubled funding, even though it knew the Afghan Supreme Court was not interested in funding or otherwise sustaining those activities" (SIGAR 2015b: 21). Programming never meaningfully engaged with the long-standing pillars of legitimacy in Afghanistan: religion, cultural norms, and provision of public goods. Instead, the focus was on public relations. RLS-Formal sought to improve the judiciary's image based on the mistaken idea that the potential users lacked sufficient information about state courts, but Afghans already recognized that courts were not credible dispute resolution forums. Neither judicial performance nor public perception of the judiciary improved (De Lauri 2013; SIGAR 2015b: 19; Singh 2015). Avoidance of the courts was entirely rational. Even if the quality of justice had improved through training or administrative reform, nothing in the program addressed the judicial system's inability to enforce its judgments or prevent outside interference.

RLS-Informal

By 2010 policymakers increasingly viewed non-state justice as a more productive avenue for international engagement. This belief was the animating idea behind RLS-Informal. The nearly $40 million Checchi-administered RLS-Informal program operated from 2010 to 2014. RLS-Informal emphasized "access to fair, transparent, and accountable justice for men, women, and children by (1) improving and strengthening the traditional dispute resolution system, (2) bolstering collaboration between the informal and formal justice systems, and (3) supporting cooperation for the resolution

[13] The Afghan government's lack of interest in meaningful reforms was not only foreseeable; it was foreseen. The 2012–2014 RLS-Formal Performance Monitoring Plan noted there was "High" risk of "resistance to training activities" and "lack of capacity to absorb program technical assistance" (Tetra Tech DPK 2012: 5). It also identified a "Medium-High" risk regarding "cooperation and productive relations with counterparts" (Tetra Tech DPK 2012: 5).

of longstanding disputes" (Checchi and Company Consulting 2014b: 1). It sought to improve the quality of non-state justice and to strengthen linkages between the state and non-state justice systems. The program's more pressing goal, however, was to supplement and consolidate U.S.-led counterinsurgency efforts. RLS-Informal envisioned filling the supposed "justice vacuum" in territories that had "been 'cleared and held' by the military and the Taliban 'courts' removed" (Dunn, Chisholm et al. 2011: 17).[14] Localized non-state justice mechanisms were to be strengthened, or in some instances created, to prevent " 'teetering' areas from reverting back to Taliban justice and influence" (Dunn, Chisholm et al. 2011: 17). RLS-Informal engaged with non-state judicial actors in targeted areas. Yet, the program largely echoed work done for state actors and drew on the same suspect template. Capacity-building was a major focus of elders and other informal-justice actors. Training programs addressed state and sharia law as well as administrative processes. Beyond training, RLS-Informal sought to "encourag[e] women's participation in TDR [traditional dispute resolution] processes, implement[] a program of public outreach, and build[] networks of elders" (Checchi and Company Consulting 2012b: 1).

RLS-Informal attempted to engage with the long-standing pillars of legal legitimacy by working within the tribal system. It sought to capitalize on the perceived desire of many local actors to increase their social standing (Checchi and Company Consulting 2014a; Schueth, Naim et al. 2014). From the donors' perspective, this dynamic could allow them to more effectively provide order. The external evaluation does identify some modest changes in response to outside assistance (Schueth, Naim et al. 2014), but local justice actors certainly did not change their approach to state justice. Although RLS-Informal sought to increase linkages between state and non-state justice, the flow was one-directional. More cases were referred to the non-state system, but referrals to state courts did not increase (Schueth, Naim et al. 2014: 32–34). This dynamic reflected Afghans' continued low regard for state courts.

Strengthening non-state justice raised even more logistical issues than strengthening the state system. A judge may be corrupt or unwilling to engage program implementers, but at least it is clear who is and is not a judge. When engaging tribal or religious authorities, participant selection was an inherently fraught process. Program staff genuinely attempted to understand each

[14] The United States Institute of Peace implemented a smaller State Department–funded program that sought to bring stability by bolstering non-state justice actors' capacity, which faced challenges similar to those encountered by RLS-Informal (Wimpelmann 2013).

locality. Nevertheless, grasping local dynamics and how international assistance would subsequently influence communal relations far exceeded their technical capacity. Although reports cast local participants as driven solely by a desire to help the broader community, often they were more motivated by the ability of international assistance to enhance their influence (Coburn 2013). Equally troubling, the program lacked a clear, coherent approach for navigating the tensions between state law, including international human-rights norms, and sharia law. Program decisions were inevitably somewhat ad hoc and based on short-term calculations. While RLS-Informal tried to understand local realities, insurgents almost invariably knew areas better. International intervention occasionally even made local tensions worse and resolution of long-standing disputes more difficult (Coburn 2013: 37).

Counterinsurgency was the U.S. government's top priority. RLS-Informal's own performance management plan explicitly states that RLS-Informal aimed to "help eliminate Taliban justice and defeat the insurgency" (Checchi and Company 2013: 3). RLS-Informal envisioned strengthening the non-state justice sector to undercut the Taliban's justice system. Simple enough in theory, but it proved deeply challenging and occasionally counterproductive. First, RLS-Informal laid the groundwork for counterinsurgency efforts on the shaky foundation of the local authorities most willing to collaborate with the United States rather than the most influential religious and tribal authorities. Therefore, program achievements depended on outside support and quickly evaporated once that support was gone. Second, RLS-Informal's short-term approach generated pressure for quick wins and demonstrable metrics, but even these were often in short supply (Checchi and Company 2014a). Third, the approach incorrectly assumed the existence of a justice vacuum in which new dispute-resolution efforts could thrive. If there were truly a void, then RLS-Informal's efforts to strengthen or create new non-state dispute resolution mechanisms would have made sense. In reality, tribal structure, warlords, or the Taliban underpinned order at the local level. RLS-Informal's favored local representatives were often not the most prominent community members. Whatever authority these individuals had largely reflected international assistance rather than local standing. Shuras set up by international actors could destabilize the situation by distributing large amounts of external funding as well as empowering individuals without substantial popular support. Interviewers who worked on internationally funded non-state justice programs even suggested that some individuals with access to international support used aid and the threat of U.S. military force to pursue personal agendas and vendettas (International NGO Professional 2014b; NGO Legal Specialist 2014).

Countering the Taliban required far more than providing an alternative justice venue. In many cases Taliban justice was comparatively well regarded, particularly in the Pashtun heartland (Kilcullen 2011: 35–45; Weigand 2017). The Taliban actively sought to prevent local communities from using state courts or U.S.-backed dispute resolution forums, even issuing "night letters" that threatened to kill all individuals and the entire families of those who collaborated with the international programs or the Afghan government. Even when villages were under the protection of international military forces, the Taliban promised to retaliate once those troops left (Johnson, DuPee, and Shaaker 2017). RLS-Informal could do little to counter the Taliban's credible threats of both immediate and long-term violence.

USAID Initiatives: Implementation versus Program Design

Justice-sector program implementation in Afghanistan faced severe challenges, but it also spoke to more fundamental problems with USAID management and oversight. Invariably, USAID sought to control the scope and content of its projects as well as their implementation.[15] This tendency to micromanage was reflected in contracts that "specif[ied] precisely what the U.S. contracting partners are to do at every step of the way throughout a project" (Carothers 2009: 26).[16] The level of centralized control in the Afghanistan contracts was stunning, particularly given the fluidity of the situation and the need for a nuanced, localized approach when trying to engage local communities. In addition to USAID's inherent contractual power to modify programming, Tetra Tech and Checchi were required to submit elaborate performance-monitoring plans for approval. USAID's drive toward ever-greater quantification and control stemmed largely from the need to establish to the U.S. Congress exactly how funding was spent and what had been achieved. Yet, USAID's in-house capacities had been largely "hollowed out" (Carothers 2009: 20). This vicious cycle was particularly intense in Afghanistan because of the extensive demands of contracting generated a general shortage of qualified applicants.

[15] Grants to implementing partners, often structured under "co-operative agreements" rather contracts, were awarded to NGOs rather than to for-profit contractors. However, USAID still retained extensive control over non-profit implementers.

[16] Stanger (2009: 109–135) provides an overview of the rise of for-profit USAID contractors and its effect on U.S. international assistance.

Justice Sector Support Program, 2005–2014

The U.S. Department of State's Bureau of International Narcotics and Law Enforcement Affairs (INL) served as the lead agency tasked with coordinating U.S. rule-of-law assistance in Afghanistan. Largely implemented by Pacific Architects and Engineers, the State Department's Justice Sector Support Program (JSSP) was the "primary capacity building vehicle" for judicial state-building (U.S. State Department 2015b). The JSSP reflected grand ambitions with $241 million in expenditures from 2005 to 2014. It was the largest single rule-of-law program in Afghanistan. JSSP featured three major components. The first focused on training for regional justice-sector officials, such as judges, prosecutors, and defense attorneys. The second concentrated on establishing a case-management system. The third component emphasized building the administrative and technical capacity of relevant ministries.[17]

The JSSP offered training, capacity building, and technical assistance to the attorney general's Office, the MOJ, the Ministry of Women's Affairs, the Interior Ministry, the Independent National Legal Training Center, and the Supreme Court. Staff numbers were significant. In 2011, for example, the JSSP employed "93 Afghan legal experts, 65 American advisors, and over 100 Afghan support staff" (U.S. Senate Foreign Relations Committee 2011: 38). Training was undertaken on a massive scale, with "a grand total of over 300 JSSP courses training over 13,500 students" (SIGAR 2014b: 2). Training occurred in all thirty-four provinces for judges, prosecutors, defense attorneys, and other legal personnel. By the end of 2014 the JSSP had developed and implemented a comprehensive case-management system active in eighteen provinces, featuring data on 104,000 cases (SIGAR 2015a: 137). The program assisted with legislative drafting and law implementation. It also created new institutions—notably, the attorney general's Anti-Corruption Unit, the Afghanistan Independent Bar Association, and the MOJ's Planning Directorate.

Despite the JSSP's broad reach and substantial funding, at its core the JSSP was a straightforward judicial capacity-building program that invested heavily in the state justice sector. Although larger in scope, both the means and the ends of the JSSP were strikingly similar to those of USAID's AROLP

[17] The International Development Law Organization, rather than Pacific Architects and Engineers, became the implementer of regional training activities under the first component of JSSP in January 2013.

and RLS-Formal programs. In terms of results, comprehensive audits of the JSSP have shown no demonstrable evidence that the program advanced the rule of law or even met its own programmatic objectives (SIGAR 2014b, 2015b). Despite extensive training, no key justice-sector institutions displayed a meaningful commitment to uniform application of the law or a willingness to follow it. New laws were passed but not consistently enforced. All major justice-sector institutions remained firmly under the control of "one of the most corrupt regimes on the planet" (Paris 2013: 538). The State Department cast the Afghan state's failure to address corruption as a failure of "political will," rather recognizing that corruption was intrinsic to the system.[18] The JSSP consistently faced major management, oversight, and implementation issues, including a series of "poorly designed deliverables," which in turn led to the actual deliverables produced being "useless" (SIGAR 2014a: 5).

DOD Rule of Law Field Force-Afghanistan (ROLFF-A), 2010–2014

The U.S. military became directly involved in efforts to promote the rule of law in Afghanistan through the Rule of Law Field Force–Afghanistan (ROLFF-A) initiative under the auspices of the Combined Joint Interagency Task Force–435. ROLFF-A operated from September 2010 to February 2014. The Department of Defense spent approximately $24 million subsidizing the state justice sector. ROLFF-A sought to "provide essential field capabilities and security to Afghan, coalition and civil-military rule-of-law project teams in non-permissive areas, in order to build Afghan criminal justice capacity and promote the legitimacy of the Afghan government" (Department of Defense Central Command 2010). The program aimed to (1) enhance human resources, (2) construct justice infrastructure, (3) increase public awareness and access to state courts, and (4) improve physical security for judges and other judicial actors in ten provinces (Department of Defense 2012b: 75).

Apart from the fourth priority, ROLFF-A's objectives were decidedly conventional. ROLFF-A attempted to strengthen the state justice system by working primarily with the Supreme Court, the Ministries of Justice and of the Interior, and the High Office of Oversight for Anti-Corruption. The program proclaimed neutrality regarding non-state justice, "provided that

[18] For instance, according to a SIGAR report, "State said the Attorney General's Office (AGO) lacks the political will to prosecute high-level, corrupt officials" (SIGAR 2013a: 135).

dispute resolution is not administered by the Taliban or other insurgent groups" (Department of Defense 2012a: 113). Although the national government had yet to formalize its relationship with non-state justice actors, the program sought to engage "Afghan Justice Sector actors to build linkages between the two systems" (Department of Defense 2012a: 113).

ROLFF-A was fully integrated within larger stability and counterinsurgency efforts in Afghanistan that stressed that "establishing the rule of law is a key goal and end state" (Amos and Petraeus 2008: 360). As the program's commander, Brigadier General Mark Martins, explained, ROLFF-A sought to "establish rule-of-law green zones" (Martins 2011: 25). Over time a "hub-and-spoke linkage between green zones in key provinces and districts" would emerge, which in turn would help "create a system of justice at the subnational level" (Martins 2011: 27). In other words, the rule of law was conceptualized as a cornerstone of broader efforts to build the state from the bottom up, while failing to acknowledge just how ambitious, optimistic, and ahistorical this belief was.

ROLFF-A faced a multitude of serious issues. Its programming incorrectly assumed that stabilization and rule-of-law programming were functional equivalents. As with the RLS-Informal initiative, ROLFF-A sought to build stability on quick wins and demonstrable outputs to fill a nonexistent justice vacuum. The program's approach was at odds with a legal landscape defined by fierce, ongoing combat, and its achievements remain speculative. The Defense Department does not even know how much money was spent (SIGAR 2015b). Ironically for a military initiative, security failures undermined the program. U.S. military forces generally could not protect against Taliban attacks in the short term, let alone once programming ceased. The mere presence of the military could incite violence as local actors scrambled to acquire external resources (Baczko 2016).

The overt militarization of rule-of-law assistance and its counterinsurgent use in Afghanistan generated additional problems. First, the assumptions underpinning counterinsurgency operations were dubious. Little evidence exists that aid actually advanced security and stabilization goals in Afghanistan or more generally (Böhnke and Zürcher 2013; Zürcher 2017). The strategy reflected a unilateral decision by U.S. policy makers rather than a cooperative strategy with the Afghan regime. Afghan government officials desired military support to defeat the Taliban but were decidedly uninterested in furthering legal reform. After all, why would non-state actors suddenly embrace a state legal system that showed no sign of improving and at the cost of potential insurgent retaliation? The U.S. military interventions never grappled with the fact that a hub-and-spoke system made little sense if the state justice sector failed to simultaneously improve.

Second, the program reflected some notable misunderstandings and misperceptions. The Defense Department consistently stressed that the non-state actors it engaged were authentic, traditional, and organic. In reality, "authority over life and death was simultaneously located in other institutions and actors, namely external interveners," rather than non-state justice actors, who largely depended on external forces for their authority (Wimpelmann 2013: 419). Moreover, the enforcement of "rule-of-law green zones" within a hub-and-spoke system relied on the prospect of external force. State courts had little ability to implement their rulings. Even more troubling, the thoroughly corrupt, deeply compromised state judicial institutions could not provide stability. Even if all areas were cleared and every insurgent defeated, the state would still lack a legitimate legal order. There was no hub to build spokes around. ROLFF-A starkly highlights external actors' limited capacity to advance the rule of law even when it is overtly backed by military force. Even in the best of circumstances, establishing the rule of law takes decades. Thus, a rule-of-law green zone means little unless it exists for years and is linked to powerful domestic constituencies committed to the rule of law.

The United Nations

The United Nations had been involved in Afghanistan since the 2002 Bonn Accords negotiations. The United Nations Assistance Mission in Afghanistan (UNAMA) was established in March 2002 (UNSC 2002b).[19] UNAMA supported programs in various areas, including electoral administration, provision and coordination of foreign aid, and efforts to increase security.[20] Under UNAMA, there was a Rule of Law unit and a Human Rights unit. The Rule of Law unit sought "to establish a fair, impartial, accessible and accountable judicial system" through technical support, advocacy, and research (UNAMA 2015b). UNAMA's Rule of Law unit had chaired the Criminal Law Review Working Group (CLRWG), a major coordinated effort among international donors and national stakeholders to revise the country's criminal

[19] UNAMA's mandate was for one year, but it was renewed annually.

[20] For an overview of the United Nations' work in Afghanistan, see Saikal (2012).

codes, since 2009. For example, the criminal procedure code was revised and passed in 2014 (International Rule of Law Professional 2015). UNAMA's Rule of Law unit also chaired the Rule of Law Board of Donors forum, which tried to coordinate rule-of-law promotion efforts led by major donors and program implementers. The Human Rights unit "analyzes and reports on the human rights situation in Afghanistan and engages in protection, advocacy and capacity-building activities" (UNAMA 2015a). These units worked both independently and in partnership with state institutions. The U.N. mission, however, was rarely a major player in the justice sector, at least in terms of influence or resource allocation.

UNDP

UNDP launched its first major justice sector project, called Rebuilding the Justice Sector of Afghanistan, in 2003; it lasted until 2005.[21] As the name implies, it focused on the state justice sector, as did its successor projects: Strengthening the Justice System of Afghanistan (SJSA) (2006–2009) and Justice and Human Rights in Afghanistan (JHRA) (2009–2015).[22] Although programming evolved, it consistently emphasized the three most prominent justice sector institutions: the MOJ, the Attorney General's Office, and the Supreme Court (see, e.g., UNDP 2005, 2009b, 2012). Assistance was also provided to secondary institutions from 2003, including the Judicial Reform Commission, the National Judicial Training Centre, and the Afghanistan Independent Human Rights Commission. Aid emphasized standard capacity-building activities such as advice, training, and material assistance. For educational institutions, aid focused on developing materials and curriculum alongside technical assistance (Tondini 2010: 57–58; Rawkins and Hashmi 2014). Programming also supported the creation

[21] The UNDP, however, never implemented programs as large as the major U.S. initiatives. For instance, the JHRA Phase 1 from 2009 to 2012 was a US$12 million program (UNDP 2012: 25). Coinciding with the SJSA was the Access to Justice at the District Level (AJDL) Project. This project focused on the helping state courts project their authority more effectively at the district level. AJDL sought simultaneously to strengthen citizens' capacity to claim their rights and enhance justice personnel's ability to "deliver justice in compliance with the rule of law and human rights standards through capacity building, technical assistance, and provision of infrastructure" (UNDP 2009a: 11). AJDL activities were subsumed by the JHRA in 2009. The UN and UNDP also implemented the Provincial Justice Coordination Mechanism Project to try to improve donor coordination in the provinces (UNDP 2009a: 11).

[22] The JHRA programme had three phases: (1) 2009–2012, (2) 2012–2014, and (3) 2014–2015 (Rawkins and Hashmi 2014: 9). This assessment covers through the end of phase 2.

of judicial infrastructure. In the JHRA's later incarnation, legislative drafting and reform were prioritized, with an emphasis on ensuring formal compliance with "international standards as set out in international conventions to which Afghanistan is a signatory" (Rawkins and Kamawi 2012: 27).

UNDP made some modest contributions, and its work hewed closely to the official rule-of-law development plans. It conducted a large number of training programs for judges, legal personnel, state employees, and law students, as well as community and religious leaders (Tondini 2010: 57; UNDP 2011b: 66–67). UNDP established a unit within the MOJ to review legislation for consistency with international standards, which UNDP argues made "improvements to several pieces of legislation including the Civil Code, the Criminal Procedure Code, the Child Act, as well as the Law on Elimination of Violence against Women" (UNDP 2013c: 30). At the government's behest, UNDP funded the construction of much-needed judicial infrastructure outside the capital. Yet even here the results were mixed. Citing a litany of management, staffing, security, and land rights issues, program implementers frankly admitted, "UNDP lacks the professional and technical capacity to undertake infrastructure work directly" and advised against future building projects without major design and implementation changes and a notable increase in the capacity of UNDP Afghanistan (UNDP 2012: 24).

UNDP programming faced serious issues that dramatically undermined its impact. These included major security challenges, which were especially pronounced outside the capital. While management issues are commonplace in foreign assistance, the scope of UNDP's issues was staggering (see, e.g., Rawkins and Hashmi 2014: 20–38). UNDP collaborated with state institutions. Thus, its work hinged almost entirely on the goodwill of its state counterparts, which at the highest levels remained uncommitted to rule-of-law ideals. As with the other programs discussed, the limits of change that can be achieved by donors without state support are stark. Although they ensure Afghanistan's formal statutory compliance with its international obligations, legislative changes make little difference when state institutions cannot or will not follow the law.

Programs depended on high-level officials, so when they questioned UNDP initiatives it caused serious concern. In 2011 evaluators noted that all three of its main project partners, particularly the MOJ, viewed the program as an "external" project "distant physically and in terms of direct management and involvement," which substantially undermined program effectiveness (Rodriguez and Anwari 2011: 35). UNDP worked hard to improve relations with the MOJ and secured buy-in from "senior figures in the ministry, including the Minister" (Rawkins and Kamawi 2012: 13). JHRA activities were subsequently "viewed as MOJ activities supported by UNDP, rather than as UNDP activities taking place in the Ministry" (Rawkins and Kamawi 2012: 13). The program was lauded as a model of constructive cooperation (Rawkins and Kamawi 2012: 13; UNDP 2012). However, in May 2014 the acting minister of justice issued a formal complaint about the project and program staff, and "many managers and staff [were] instructed to cease their cooperation with the project" (Rawkins and Hashmi 2014: 13). Given the fluidity of Afghanistan's political situation and security environment, even modest programmatic achievements could quickly evaporate.

Programmatic Progress and Challenges

Although the programs detailed above achieved some tactical gains and helped build some state judicial infrastructure, no international program meaningfully advanced the rule of law in Afghanistan. Each faced significant challenges in a complex, legally pluralist environment with serious security concerns. The biggest problem, however, was that key state actors were not committed to the rule of law. State institutions were open to financial support or infrastructure funding, but there was little demand for the changes international assistance was supposed to facilitate. Without high-level state efforts to reduce corruption, improve judicial performance, and engage constructively with non-state tribal and religious actors, judicial state-building assistance would always achieve little regardless of expenditure or program design. Indeed, when even tentative efforts were made to investigate corrupt practices, international officials "were rebuked by Karzai's officials for misunderstanding the nature of patronage networks that served to support the government" (Chaudhuri and Farrell 2011: 285). Yet, U.S. policy makers chose to keep investing in efforts they knew were not working. Policy makers, donors, and implementers embraced wildly optimistic assumptions to rationalize the perpetuation of failing programs and policies. International, and particularly U.S., decision makers made clear policy and programmatic choices that had profound consequences.

Conceptualizing International Judicial State-Building in Afghanistan

When the Taliban regime collapsed, Afghanistan's legal system displayed a high degree of legal pluralism. Yet, most disputes were still settled based on tribal codes. Initial security arrangements outside the capital were predicated on alliances between the state and armed strongmen. International support reflected the optimistic and ultimately unrealistic idea that these tacit alliances would become less necessary as state courts gained capacity and authority. Italy, when it was the lead nation, the United Nations, and the UNDP all consistently focused on subsidization of the state justice sector. In contrast, the U.S. approach evolved over time. While aid focused on bolstering state justice-sector institutions for more than half a decade during the George W. Bush administration, Obama administration policy makers subsequently determined that subsidization alone was insufficient.

In 2009 U.S. policy rhetoric regarding Afghanistan's legal system began to shift, but its initiatives remained focused on a litany of highly repetitive training programs and capacity-building projects. Policy reflected a combination of political expediency and security concerns that bore little resemblance to the grandiose claims being made about ensuring the rule of law and promoting access to justice for ordinary Afghans. The United States consistently revamped its official approach, but improvement remained elusive. Although programs were problematic, the overarching issue was strategy. International actors attempted to build the rule of law in partnership with entities that did not want it. The Afghan regime very interested in using the law to bolster its authority but had no interest in being bound by the law itself, ending impunity, or promoting a democratic state bound by the rule of law.

Strategies for Addressing the Non-State Justice Sector

The previous section highlighted the most important programmatic initiatives undertaken with a goal of advancing the rule of law in Afghanistan. This section focuses on the strategies employed by international policy makers to further this objective. As explained earlier, there are five main strategies for conceptualizing engagement between the state and non-state systems in settings with a high level of legal pluralism such as

Afghanistan: (1) bridging, (2) harmonization, (3) incorporation, (4) subsidization, and (5) repression. These strategies are conceptually distinct though by no means mutually exclusive. Success can never be guaranteed by strategy alone, but certain environments favor certain strategies. Italy, the United Nations, and the UNDP only utilized subsidization. The United States initially focused solely on subsidization but eventually embraced a range of approaches including harmonization, incorporation, bridging and repression. Consequently, the next section focuses on how U.S. assistance evolved over time. In the end, no international actor and no particular strategy or combination of strategies was able to advance the rule of law in Afghanistan.

Subsidization Strategy, 2002–20008

International rule-of-law promotion efforts in Afghanistan began in earnest in 2003, and they focused overwhelmingly on subsidization (Gaston and Jensen 2016). The vast majority of the country's judicial infrastructure had been destroyed during the conflict. Qualified legal professionals were scarce, and their training needs were daunting (Armytage 2007; Swenson and Sugarman 2011). International assistance was a prerequisite for enabling most justice-sector organizations to function at all. Subsidization thus reflected a clear, compelling rationale: the state justice sector desperately needed improved human resources, training, supplies, and infrastructure. International aid was focused on drafting legislation and building modern state institutions compatible with the rule of law. The subsidization approach was deeply embedded in nearly all judicial state-building efforts, regardless of the funder. The 2005–2010 USAID–Afghanistan Mission's strategic objectives explicitly endorsed subsidization. It focused upon "build[ing] capacity of the formal justice sector" by (1) "decreas[ing] obstacles to citizens accessing the formal justice sector," (2) "increas[ing] professionalism of judicial sector personnel," and (3) "strengthen[ing] the institutional capacity for lawmaking and technical drafting" (USAID 2005: 17). Notably, subsidization remained the key strategy during and after the lead-nation approach (London Conference on Afghanistan 2006: 8; Tondini 2007).

International efforts initially were criticized for allocating insufficient resources to state reconstruction (Paris 2004: 226–227, Barfield 2010: 315–320). Yet, Afghanistan received an immense amount of support relative to other contemporary peace-building missions, and its aid absorptive capacity was limited (Suhrke 2011: 119–141). Merely increasing assistance does not guarantee improvement and often produces significant problems. As discussed earlier, the rule of law takes decades to establish, and there is no area where simply increasing rule-of-law assistance would have clearly translated into better justice or a state more committed to the rule of law.

Subsidization achieved little given its slow start, lack of state interest, and poor strategic choices. These choices included failing to establish nationwide security during a period of relative calm, outsourcing security to warlords, and allowing consolidation of authority in the executive and a culture of corruption to flourish. State officials cared little for internationally backed legal modernization plans. Moreover, these initiatives were decidedly uninspired. As Carothers has observed, assistance often takes a "breathtakingly mechanistic approach to rule-of-law development," one focused on replicating "institutional endpoints"—even though this approach has been widely discredited (Carothers 2003: 9). The pattern described by Carothers neatly captures Afghanistan's situation, where programmers assessed

> in what ways selected institutions do not resemble their counterparts in countries that donors believe embody successful rule of law—and then attempting to modify or reshape those institutions to fit the desired model. If a court lacks access to legal materials, then those legal materials should be provided. If case management in the courts is dysfunctional, it should be brought up to Western standards. If a criminal procedure law lacks adequate protections for detainees, it should be rewritten. The basic idea is that if the institutions can be changed to fit the models, the rule of law will emerge. (Carothers 2003: 9)

Outside investments in the state justice sector, however, produced no discernible progress toward furthering the rule of law in Afghanistan.

In 2006 international actors began to express growing interest in Afghanistan's non-state justice sector. Policy suggestions included bridging through the creation of mutually constitutive institutional links between state and non-state dispute resolution mechanisms, harmonization through attempts to ensure that non-state dispute resolution forums acted in a manner consistent with state law, and incorporation efforts to establish some sort of overarching system wherein jirgas and shuras would function similarly to courts of first instance (Bassiouni and Rothenberg 2007; Center for Policy and Human Development and UNDP 2007; Checchi and Company

Consulting 2009b). Although it was a growing area of interest intellectually, programming largely left the non-state justice sector untouched throughout the Bush administration.

A Road Not Taken

Although the United States was the most open to engaging with non-state justice, the most thoughtful and compelling attempt to constructively engage with non-state justice in Afghanistan was a major UNDP report, "Afghanistan Human Development Report 2007." The report forcefully advocated for a "hybrid model of Afghan justice" that would include:

> the creation of cost-effective ADR [alternative dispute resolution] and Human Rights Units alongside the state justice system. ADR Units would be responsible for selecting appropriate mechanisms to settle disputes outside the courtroom. This would include jirgas/shuras, Community Development Councils, and other civil society organizations. ADR mechanisms would handle minor criminal incidents and civil cases, while giving Afghans a choice to have their cases heard at the nearest state court. All serious criminal cases would fall exclusively within the jurisdiction of the formal justice system. When ADR decisions are not satisfactory to the disputants, they can be taken back to the formal, state justice system. (Center for Policy and Human Development and UNDP 2007: 12)

The report, however, provoked "an angry and threatening response from Afghan judicial and state justice institutions" (Wardak 2011: 1322). Given the opposition of key Afghan political and legal elites to sharing authority with non-state judicial actors, it is no surprise that the hybrid model gained no traction either domestically or internationally.

The UNDP report nonetheless proved influential. Ideas from the report were selectively harnessed by policy makers for numerous initiatives, including a series of U.S.-funded pilot programs, the Defense Department–backed projects that used non-state justice mechanisms for counterinsurgency discussed above, and the flawed 2010 "Draft Law on Dispute Resolution Shura and Jirga." Each initiative, however, fell short of its goals (Coburn 2013; Wimpelmann 2013). The report offered a coherent policy vision, but to be implemented, let alone work effectively, the proposed hybrid model required strong support from the Afghan justice sector and the regime more generally. There would need to be a serious commitment to building a more constructive relationship between state and non-state justice actors, as well as a broader commitment to reducing corruption and ending impunity. Absent these fundamental shifts, the hybrid model had no chance of success.

A Comprehensive but Inchoate Approach, 2009–2014

President Obama pledged to defeat the insurgency and stabilize the Afghan state while avoiding an open-ended military commitment. Nearly all U.S. aid, including rule-of-law assistance, emphasized these objectives. The top U.S. military commander in Afghanistan, General Stanley McChrystal, argued that effective counterinsurgency involved bolstering the quality of and access to both state and non-state justice mechanisms "that offer swift and fair resolution of disputes, particularly at the local level," to disrupt the Taliban and their justice system (McChrystal 2009: Sec. 2, p. 14).To achieve these goals, McChrystal believed that the international community had to "work with GIRoA [Government of the Islamic Republic of Afghanistan] to develop a clear mandate and boundaries for local informal justice systems" (McChrystal 2009: Sec. 2, p. 14).

After a comprehensive policy review, Obama ordered an "Afghanistan surge" modeled on similar efforts to stabilize Iraq against insurgent advances. The president authorized the deployment of thirty thousand additional ground troops to stabilize the country, but these forces would begin to be withdrawn after eighteen months. Obama likewise dramatically increased civilian engagement efforts, including initiatives to promote the rule of law. In addition to a vast funding increase, the administration roughly "tripled the total U.S. government civilian presence in Afghanistan from 300 to 1,000, overseeing additional thousands of contracted civilian implementing partners" (Brown, F. 2012: 3)· It also drastically increased U.S. subsidization of the state justice sector. Even after the drawdown, funding for rule-of-law programs remained well above pre-2009 levels. The administration demonstrated a willingness to try any strategy that might help defeat the Taliban insurgency, however implausible. Despite mounting evidence to the contrary, the United States remained wedded to the "vacuum" theory of judicial state-building as a foundation of the U.S. government's rule-of-law assistance approach. For instance, the 2010 "Afghanistan and Pakistan Regional Stabilization Strategy" emphasized that "justice and rule of law programs will focus on creating predictable and fair dispute resolution mechanisms to eliminate the vacuum that the Taliban have exploited with their own brutal form of justice" (Office of the Special Representative for Afghanistan and Pakistan 2010: ii).[23]

The U.S. strategy was open-minded about where Afghans sought justice, provided they eschewed Taliban courts. This approach to non-state justice displayed admirable pragmatism compared with earlier efforts, which emphasized replicating a Western-style judiciary. Yet, this agnostic approach was undermined because programs addressing the state justice system were still trying to do exactly that. With so many moving, highly contingent, and largely uncoordinated parts, the strategic objective was never clear. U.S. policy makers viewed non-state justice mechanisms as instrumentally important to their overarching goals of counterinsurgency and stabilization. There was never a serious attempt, however, to engage with the three main pillars of legitimacy: cultural affinity, Islam, and the provision of public goods, including forums for fair and equitable dispute resolution.

Simultaneous Judicial State-Building Strategies

The Obama administration's transformative plans for Afghanistan were crystallized in the unified civil-military U.S. Foreign Assistance plan for 2011–2015.[24] The plan explains:

> The principal focus of the U.S. rule of law effort is to reverse the public perception of GIRoA as weak or predatory by helping the Afghan government and local communities develop responsive and predictable dispute resolution mechanisms that offer an alternative to the Taliban shadow justice system. Assistance will be provided in support of Afghan efforts to strengthen the formal state justice system, stabilize the traditional justice system, and build a safe, secure, and humane civilian corrections system. (U.S. Mission Afghanistan 2010: 5)

[23] The idea of the justice vacuum was also important at the level of individual programming for initiatives such as RLS-Informal (Dunn, Chisholm et al. 2011: 17)

[24] My focus is on the joint plan, as it had explicit unified approval. For similar plans, see US Department of Defense and US Department of State (2009); and Office of the Special Representative for Afghanistan and Pakistan (2010).

Non-state justice was explicitly linked to counterinsurgency efforts (U.S. Mission Afghanistan 2010: 4). The United States wanted to collaborate with preexisting non-state actors but also to "re-establish[] traditional dispute mechanisms" as part of a broader effort to counter the Taliban's parallel justice system (U.S. Mission Afghanistan 2010: 5). It created non-state dispute resolution forums, albeit ones without long-standing cultural and religious roots. This endeavor involved local research on non-state systems and how they could be utilized to defeat the Taliban. Thus, each strategy ostensibly promoted goals in the legal sector, but it existed primarily to support counterinsurgency efforts.

U.S. assistance was expanded and diversified to include new strategies and engage with both state and non-state justice actors. It did not, however, produce the desired results. Behind the transformative rhetoric, the United States' bold vision for rule-of-law assistance was underpinned by four wildly optimistic "critical assumptions":

1. The Afghan government will implement its reinvigorated plans to fight corruption, with measures of progress toward greater accountability.
2. Justice and rule of law programs will focus on creating predictable and fair dispute resolution mechanisms to eliminate the vacuum that the Taliban have exploited.
3. USG [U.S. government] programs will successfully address local officials' lack of education, experience, and limited resources.
4. GIRoA action will counter obstruction from local powerbrokers whose activities are sometimes inconsistent with the Afghan constitution. (U.S. Mission Afghanistan 2010: 10)

Each of these assumptions was clearly unrealistic. The Karzai administration never sought to eliminate state corruption. The justice system displayed few signs of improvement, lacked a commitment to the rule of law, and remained subject to executive influence. As evidenced by the serious challenges faced by well-funded U.S. programs aimed at Afghan state institutions that placed a major emphasis on capacity-building, resource restraints and lack of knowledge were rarely the reason rule-of-law initiatives underperformed. Moreover, the Karzai regime and the U.S. government were in league with powerbrokers who actively opposed advancing the rule of law. Although they found warlords unsavory, U.S. policy makers partnered with them to try to defeat the Taliban, and there was never a serious attempt at disarmament. Afghanistan's rule-of-law situation saw very little improvement by the time Afghans elected a new president in 2014, in a process marked by even greater fraud and controversy than had been seen in the 2009 elections.

Subsidization

In 2009 the U.S. government renewed its pledge to "support capacity development of the formal state courts" (U.S. Mission Afghanistan 2010: 5). The new strategy's core was, in reality, a supersized reincarnation of the preexisting subsidization strategy. For example, funding from the State Department's INL Bureau for rule-of-law assistance in Afghanistan ballooned from $26.5 million in 2006 to $328 million in 2010 (Wyler and Katzman 2010: 27). Yet, massive funding increases failed to transform the Afghan justice sector or even produce notable improvements. Instead, they demonstrated the limits of what subsidization could achieve absent an ideological commitment by the state to the rule of law. As noted previously, the Karzai regime was hostile toward efforts to strengthen the rule of law, which could undermine both its freedom of action and its patronage system. The Supreme Court and other key judicial organs were open to receiving aid, but only on their terms and never in a way that ultimately threatened the overarching system.

Harmonization

A harmonization push sought to make the non-state justice system in Afghanistan operate on principles akin to the state system or, more accurately, on an idealized conceptualization of state justice that protects human rights and upholds the rule of law. It involved both supply-side and demand-side activities. On the supply side, "tribal elders/religious leaders who conduct shuras would receive training on relevant state and religious law" (U.S. Mission Afghanistan 2010: 4). On the demand side, U.S. assistance would increase Afghans' awareness regarding their legal rights and how to assert them. Assistance would thus help align the behavior of state and non-state actors and bolster non-state actors' willingness to support the state. There is little to suggest that U.S. initiatives caused Afghans to view state law more favorably or non-state justice mechanisms to operate more procedurally or substantively like state courts. Neither was there any increase in the "enforcement of the rights of women and other traditionally marginalized groups" (Wyler and Katzman 2010: 27). Harmonization failed because non-state justice actors largely remained wary of state courts, and human-rights concerns proved readily expendable when they were seen as clashing with security and counterinsurgency goals.

Bridging

Bridging between the state and non-state justice sectors in Afghanistan became a more prominent strategy after 2009. The United States sought to "establish[] linkages, as appropriate, between the informal and state systems" (Wyler and Katzman 2010: : 5). This strategy envisioned Afghan citizens enjoying free access to both systems as appropriate, predicated on a sensible jurisdictional divide. For example, whereas a judge might refer a property theft to a local jirga, state courts would retain exclusive jurisdiction over major crimes such as murder. In reality, however, the state lacked the capacity to compel most non-state justice actors to use state courts. The quality of state justice remained poor, so there was no demand for it; RLS-Informal witnessed no discernible increase in popular demand for state courts (Checchi and Company Consulting 2014a). Although aid likely increased public awareness and perhaps even access, it was a proverbial bridge to nowhere because the public demand for state justice remained scant.

Incorporation

Incorporation envisioned a partnership among the implementers of U.S. rule-of-law programs, the Afghan MOJ, and other relevant agencies "to formalize links between the two systems [state and non-state justice] to maximize the benefits of both systems and to reduce the weaknesses" (Checchi and Company Consulting 2014a: 4). In the 2009 draft, non-state justice policy was promulgated, backed by state representatives and international officials. The policy consensus was shallow, however. There was little consultation with tribal non-state justice authorities, and no clear vision existed as to how to address the de facto authority of warlords in the areas they controlled. The September 2010 "Draft Law on Dispute Resolution Shuras and Jirgas" imagined non-state justice actors as part of the state system. Consequently, it heavily regulated their jurisdiction, operations, and decision-making as well as their relationship to state courts (Ministry of Justice 2010). As such, the draft law included harmonization and bridging elements alongside incorporation. With international backing, the law staunchly asserted the state's authority to control and regulate all aspects of non-state dispute resolution. In addition, it stipulated that jirga participants "and parties of dispute shall be duty bound to observe provisions of this law" or face criminal charges, even though jirgas were only empowered to hear civil disputes and petty juvenile crimes on referral (Ministry of Justice 2010). Jirgas and shuras could not "make decisions that violate human rights of parties in dispute, especially of women and children," and all judgments would have been subject to appeal to state courts (Ministry of Justice 2010). The law was not passed, however, owing to opposition from the Ministry of Women's Affairs and the Human Rights Commission. The state was open to engagement with the non-state justice sectors, but only on its own stark terms. The central government's internal divisions ultimately prevented the law from being promulgated despite international interest in doing so. Even if a law had been passed, it would have beem unlikely to meaningfully alter how dispute resolution operated in many areas, given the weakness of state authority.

Repression

Afghan state and international forces sought to shake and undermine the Taliban justice system. At the same time, these actors realized that the popular appeal of Taliban justice constituted a profound challenge to state authority. Although rule-of-law programs were not trying to destroy the Taliban system and eliminate its core personnel, with the partial exception of ROLFF-A, they did aim to undermine the system's appeal. Doing so proved remarkably difficult, however, because the state system remained unappealing. By the end of 2014 it was clear that Taliban justice would remain an enduring feature of the Afghan legal landscape.

Conclusion

Because international assistance is invariably mediated through the state institutions and officials, Karzai's regime bears the majority of responsibility for the lack of progress toward the rule of law and the increasingly tense relationship between the state and non-state justice sectors. Yet, the international community is far from blameless. It allowed the initial outsourcing of security to warlords, and it acquiesced, and on occasion even actively supported, numerous institutional and policy decisions that were counterproductive to the goal of establishing democracy or the rule of law. International aid was always suboptimal. Judicial assistance was minimal during the crucial early interval and then focused on unneeded and unwanted legislative reforms. Once funding began to increase dramatically, no plausible strategy emerged for how the assistance would actually achieve progress toward the rule of law.

These huge expenditures of aid served to prop up fundamentally compromised state institutions; little thought was given to the constructive engagement of the tribal and religious actors who had long formed the crucial pillars of legal legitimacy. Engagement with the non-state justice sector was little better. International actors accepted the regime's increasing dependence on warlords who lacked any interest in the rule of law. Over time, the Taliban and its justice system grew in strength and popular legitimacy, owing in large part to the Afghan state's corruption and reliance on warlords. International interest in engagement with the non-state justice, however, emphasized counterinsurgency rather than justice. International assistance ultimately produced very few benefits and could even be counterproductive. The regime constructed by President Karzai and his allies over the years was a clear limiting factor, but these outcomes were not inevitable. However, as highlighted by the structured comparison of Timor-Leste and Afghanistan in Chapter 8, the international community made a series of policy choices that had real consequences for the trajectory of judicial state-building in Afghanistan.

8
Conclusion

Introduction

This chapter explores how overarching law and governance structures influence the nature of legal pluralism and how, in turn, attempts to establish the rule of law in legal pluralist environments can influence and shape efforts to establish a democratic state bound by the rule of law. The first section, by drawing on insights garnered from the case studies of Timor-Leste and Afghanistan, provides an overview of theory building and theory testing. The second highlights lessons and implications for international and domestic actors. The third section identifies areas for further research. These are followed by a general conclusion.

Overview

Together, the case studies offer examples of both theory-building (Timor-Leste) and theory-testing (Afghanistan) (George and Bennett 2005: 114–120) to generate insights into post-conflict judicial state-building in legally pluralist environments. More specifically, I argue that Timor-Leste was successful because: (1) despite fierce political competition, all major political actors and parties remained committed to constructing a democratic state underpinned by the rule of law; (2) there was a credible and sustained effort to develop institutions that promote democratic accountability, inclusiveness, and the rule of law; and (3) the state meaningfully engaged and collaborated with key non-state actors through vibrant, competitive local suco elections and also developed strategies to make it easier and advantageous for suco chiefs to work with the state rather than against it. While the process was imperfect, Timorese state officials also mediated between the international community and local-level figures. This helped to transform a competitive legal pluralist environment into a cooperative one.

Contending Orders. Geoffrey Swenson, Oxford University Press. © Oxford University Press 2022.
DOI: 10.1093/oso/9780197530429.003.0008

While the state served as the primary mediating force, the international community also played a constructive role. In the theory-building case of Timor-Leste, the state made progress toward a modern democratic state bound by the rule of law through competitive national elections and by laying the groundwork for professional and independent judicial institutions. Equally important, the state's compelling vision of a democratic state committed to the rule of law combined with competitive, local suco elections established a cooperative relationship with the non-state justice sector. The international community generally reinforced positive trends by subsidizing the nascent state justice sector and, particularly after the 2006 crisis, working to link the state and non-state justice sectors. Despite this synergy, Timor-Leste's Lusophone legal system has presented real challenges. Going forward, far more widespread use of Tetum, the dominant local language, within the state legal system and state institutions will likely be essential to fully consolidate the rule of law. Even with these caveats, Timor-Leste is a surprising success story.

Afghanistan served as a theory-testing case. The case-study chapters highlight how the chance to build a cultural intelligible but also more inclusive and effective democratic state in Afghanistan was wasted. Indeed, it was never even seriously attempted. Despite rhetoric to the contrary, the country's post-Taliban regime never displayed a normative commitment to the rule of law. The now-defunct Islamic Republic of Afghanistan focused on maintaining power and securing rents rather than performing state functions well. Second, the post-Bonn regime systematically undermined institutional, legal, and political checks designed to limit its authority. These efforts include suppressing political parties, manipulating elections, and undermining judicial independence. Finally, during this period the Afghan state never seriously engaged with key tribal and religious non-state justice actors, who have historically provided the building blocks for a legitimate order. As a result, tribal and religious legal authorities often felt excluded from the new regime and had little stake in its success. The regime partnered with warlords, but it was a purely transactional relationship based on the exchange of resources for support. The Afghan state failed to mediate between the international community and local actors, particularly the tribal and religious authorities that have historically proven essential for legitimate rule. While by no means the sole cause, this divergent approach toward non-state legal authorities is an important contributing factor that helps explain judicial state-building's failure in Afghanistan and the corresponding slide from competitive legal pluralism into combative legal pluralism with an increasingly potent, and ultimately triumphant, Taliban insurgency.

Immediately after the end of the first Taliban regime, the international community exercised tremendous sway over the nascent Afghan state's institutional structure and its leadership. International backing secured the top state post for Karzai and encouraged immense concentration of power in the executive. Even after the initial period, the international community, and particularly the United States, retained significant influence over the Afghan government. Afghanistan consistently ranked as one of the world's most aid-dependent states (Hogg, Nassif et al. 2013: 52). Despite this influence, international involvement did little good and was occasionally even counterproductive to promoting the rule of law. Nor did international engagement foster constructive engagement with the tribal and religious non-state actors who played and continue to play a key role in maintaining local stability and influencing the extent of popular and elite support for the state. When international actors finally engaged with non-state authorities, they did not prioritize promoting a more just legal order. Non-state actors were treated as mere cogs in a deeply flawed counterinsurgency approach.

Learning from Lessons Learned and Not Learned

Advancing the rule of law in both Afghanistan and Timor-Leste presented a daunting challenge. Yet these efforts had real promise, even in Afghanistan. In late 2001 the Taliban regime wilted away, with minimal resistance. The new, multiethnic Islamic Republic had a real opportunity to offer better, more effective governance. That opportunity was squandered because of both domestic and international choices. The Afghan state subsequently endured a prolonged legitimacy crisis and a powerful, and ultimately victorious, insurgency that promised the law and order the state failed to provide. To cast these outcomes as inevitable, however, serves to excuse bad policy and invite future disasters. Timor-Leste likewise had an opportunity to build a break with the past and establish a new state committed to the rule of law that enjoyed legitimacy among the population nationwide. Success was by no means predestined. The scale of political and economic challenges at independence was staggering. Political elites were fiercely divided, and the possibility of violence was very real; indeed, political contestation exploded into overt violence in 2006, and the upheaval shook the political and legal order to the core. But with international support, Timor-Leste's domestic leadership and institutions (both state and non-state) responded to the challenge with a consistent commitment to democracy and the rule of law. While serious challenges persist, Timor-Leste has made impressive progress.

Timor-Leste and Afghanistan are not alone. Establishing a legitimate legal order and advancing the rule of law after conflict will constitute a major domestic and foreign policy challenge for the foreseeable future. Because extensive legal pluralism characterizes many societies, particularly conflict-prone ones, future interventions and subsequent efforts to advance the rule of law will almost certainly occur in places marked by a high degree of legal pluralism, such as Yemen and Syria. As a result, Timor-Leste and Afghanistan offer insights into improving future judicial state-building endeavors.

A Stitch in Time . . .

Afghanistan's own history shows that a viable, legitimate order was possible. Before the late 1970s communist coup, the country had enjoyed decades of domestic tranquility. In the 1990s the Taliban established a monopoly on the use of force under even more chaotic conditions. After the end of Taliban rule in 2001, Afghanistan enjoyed a period of relative peace before 2006, when a full-scale insurgency developed. While Afghan society has certainly demonstrated a capacity for violence, order is also prized. This is evident from the widespread veneration of Pashtunwali and from Afghans' acceptance of a national state, albeit a limited one, for centuries. Rather than bemoan the country's hopelessness, it is important to critically examine how domestic and international postwar domestic and international state-building efforts could have been undertaken in way that constructively builds on Afghan history and culture.

International actors have their greatest influence and the broadest range of feasible options at the onset of post-conflict reconstruction (Doyle and Sambanis 2006). The state judicial sector in Afghanistan faced daunting challenges, including minimal infrastructure, serious human-resource shortfalls, dubious legitimacy, and skepticism from powerful non-state authorities. There was a clear, constructive opportunity for international assistance to help address these issues. International aid did just that in Timor-Leste. International troops helped provide security and stability in the wake of the 1999 referendum and the 2006 crisis. International professionals served temporarily in the state legal system itself and helped trained domestic legal staff. After the crisis, once the state legal system had been stabilized, international support was crucial in helping ensure that serious crimes could reach state courts and in building increased awareness in suco councils about relevant state laws. By contrast, Italy—in its role as lead nation for law and justice—took little action in Afghanistan until 2003 and achieved few successes before it was relieved of its duties in 2006. Italy also ignored non-state justice and focused exclusively on the state. Italy's performance has been rightly criticized, but the loss of this window of opportunity was not solely Italy's fault. The United States decided to topple the Taliban and was by far the most dominant international player in determining policy afterward. Italy's lead-nation status was not inevitable or even particularly logical, but rather a reflection of the U.S. policy of outsourcing to other nations as many aspects of state-building as it could during the crucial early period.

While the immediate post-conflict period provides a natural window of opportunity, it is equally important to capitalize on reset moments. Timor-Leste's top state officials had legitimacy stemming from the independence struggle, and established institutions made a good-faith effort to promote democracy, provide a just state legal order, and engage constructively with non-state justice actors. The legitimacy of the state and its new institutions meant that non-state leaders were increasingly open to the judicial state-building project. But post-conflict situations are almost invariably fragile. State-building and indeed law and order itself broke down during the 2006 crisis. The crisis was a profound shock to the political order, but ultimately both political rivals reaffirmed their commitment to democratic principles and a legal order bound by the rule of law through actions, not just rhetoric. With the notable exception of Portugal (which stayed focused exclusively on subsidization), international policy makers not only continued to develop key state institutions—they collaborated with state and non-state authorities to help bridge gaps between the state and non-state justice as well as to develop a more cooperative relationship between those systems.

Setting the Foundation

State legitimacy and institutional arrangements are vital for developing the rule of law and establishing a workable relationship between state judicial entities and their non-state counterparts. International rule-of-law assistance is almost invariably mediated through state institutions and officials, even when targeting non-state actors. It matters immensely whether the state is seen as legitimate and whether those entrusted with the state's authority care about promoting a democratic state bound by the rule of law. Providing domestic security is essential. Political space must exist for constructive engagement. With the benefit of international assistance, Timor-Leste did both by providing a credible security guarantee and a legitimate overarching state and justice system underpinned by democratic values and a good-faith effort to engage with non-state justice authorities. In contrast, President Karzai's regime was decidedly uninterested in promoting the rule of law or democratic accountability; in fact, it opposed meaningful efforts to advance either. In general, international policy makers should be deeply skeptical of the assurances of corrupt rulers with authoritarian tendencies. The international community also bears responsibility for aiding and abetting Afghanistan's governance disaster.

The rule of law is tightly linked to democratic accountability. The ability to change the government through free elections is a necessary, though not sufficient, condition to support democratization and foster the inclusive institutions that underpin the rule of law (Acemoglu and Robinson 2012). Moreover, the legal order is inexorably interconnected with a state's political institutions. In theory, democracy is not required for the rule of law. Yet, as a both practical and empirical matter, the rule of law demands democratic government in some form (Habermas 1995). While democratic rule does not inevitably produce the rule of law, it appears to be a functional prerequisite (O'Donnell 2004). Democratic governance can also help facilitate a more cooperative rather competitive relationship between state and non-state authorities by facilitating the entry of non-state groups into mainstream politics and by lowering the barriers to mobilization and accessing some degree of state-sanctioned political authority.

Voting alone is not enough. Electoral systems matter immensely, and strong evidence suggests that "democracy survives better under parliamentarism then under presidentialism" (Tsebelis 2002: 75). Granted, no electoral process is perfect. Proportional representation, however, is widely acknowledged as the most suitable system for post-conflict states because it encourages broad-based regimes and complicates efforts to consolidate authoritarian rule (Reilly 2002). This is true in general and after conflict. Joshi, for instance, found that the "survival of a post–civil war transition is likely to increase with the adoption of the parliamentary system of government by almost 76%" (Joshi 2013: 761). Proportional representation also helps ensure that the political system represents the interests of non-state actors. Thus, as stakeholders within the system, non-state justice actors are more likely to support the overall political order, or at least not actively seek to destroy it.

In the early years, the international community exercised meaningful influence over the development of the nascent Afghan state's institutional structure. International backing secured the top state post for Karzai initially and encouraged the president's concentration of power and a winner-take-all dynamic. After taking office in 2001, President Karzai behaved accordingly. He ruthlessly undermined, co-opted, or eliminated all major rivals within the state. In 2014 he ensured that his preferred successor became president in a deeply compromised election that required the United States to broker an extraconstitutional power-sharing arrangement. In Afghanistan there was far less need to actively compete for the support of non-state justice leaders, or indeed the population writ large, when fraud and corruption offered much more promising route to electoral success. By 2018 even USAID in Afghanistan itself admitted that "elections are not yet perceived by the public as an effective way to influence public policy" (USAID 2018: 55).

In contrast, international involvement in East Timor encouraged political pluralism even before Timor-Leste officially became an independent state. After independence in 2002, Timor-Leste's executive structures promoted democratization by establishing a strong prime minister and a relatively weak president. For parliamentary elections, Timor-Leste adopted a closed-list proportional-representation election system, with a 3 percent vote threshold, and encouraged broad-based strong political parties while limiting party fragmentation. A sensible electoral system, a well-constructed political parties law, open electoral competition, and well-executed elections have fostered a democratic culture in Timor-Leste. Two major political parties, Fretilin and CNRT, have led functioning governments, and credible smaller parliamentary parties have emerged. Equally important, election results have consistently been accepted by all the major parties. In 2007 a peaceful transfer of power occurred between Fretilin and Gusmão's CNRT, with the first coalition government and a strong but loyal parliamentary opposition. In each election, the major political parties sought the support of suco chiefs and other key non-state actors. After each election, the victorious parties continued to work constructively with suco authorities.

For its part, the international community deserves credit for helping create a political environment from the UNTAET era and beyond in which vibrant multi-party competition could thrive. Notably, it avoided picking winners or providing major institutional advantages to Gusmão or Fretilin, a choice that would have distorted political competition. International actors initially worked most closely with Gusmão but shifted their focus to Alkatiri after Fretilin's success in the 2001 elections. UNTAET officials did not attempt to work solely with Gusmão after the independence front-CNRT disbanded in 2001. UNTAET also supported a drafting process that required a supermajority to ratify the constitution, thus encouraging compromise and preventing any one party from dominating the process. Despite criticism that the United Nations was fostering divisions, it ultimately proved wise not to take sides or place a thumb on the scale for a particular political party. The United Nations and the broader international community were well placed to engage with whichever duly elected officials were in power at a given time and to facilitate open political competition more generally.

Initial international favoritism can have serious consequences, because institutional arrangements can prove very resistant to change once they are entrenched. Individuals who reach the top spot first have a disproportionate advantage going forward and can use their position to undermine democratic competition in the future. International actors need to be cautious with their support. They must push for free elections, vibrant political parties, and the creation of political accountability mechanisms to ensure that the people themselves have real control over who leads their country. In turn, this helps ensure that the state, and its legal system, appear legitimate to non-state legal authorities as well as the population more broadly.

The international community, and particularly the United States, viewed security as the top priority in Afghanistan after 2001. From the outset, concerns about the rule of law were relegated to the background. The international community's ability to promote the rule of law was heavily circumscribed by its entrenched commitment to the regime and an emphasis on security over justice. Karzai's administration sought to retain power and exercise authority unconstrained by law. Although international policy makers wanted a liberal democratic polity in Afghanistan that embraced the rule of law, they consistently compromised on this aspiration. The Karzai regime recognized that international—and especially U.S.—policy makers believed that the state-building project could not be seen as failing, and that to admit that it *was* failing would have risked its being seen as tacit acceptance of a Taliban victory. Over time the Afghan state's profound endemic weakness gave state officials' greater ability to operate independently from its international backers as the prospect of a Taliban victory became increasingly plausible. This dynamic was consistently reinforced throughout the Trump administration until President Biden eventually decided to remove U.S. forces.

It is tempting to believe that a focus on security somehow excuses or justifies a lack of interest in a democratic state bound by the rule of law. In reality, poorly governed conflict-prone states can present a serious domestic and international threat. Support for the rule of law can help improve security. Timor-Leste provides an illustrative example. As instability in Timor-Leste came increasingly to be seen as a security threat to Australia after the 2006 crisis, Australia responded not by forming a partnership with unsavory criminal elements or wholeheartedly backing certain leaders (and permitting them to marginalize other political groups). Instead, it redoubled its foreign assistance to promote the rule of law and backed the presence of international forces to ensure security (deployed with permission of the Timor-Leste government). While by no means perfect, Australian and other international support had a positive impact on developing the rule of law and promoting security; furthermore, and more broadly, it highlighted the possible of synergy, rather than rivalry, between these two goals.

Democracy and the rule of law demand more than free and fair elections. Electoral integrity is a prerequisite. In Timor-Leste, despite occasional violent clashes between partisans, elections have consistently been free and fair. Initially international support helped ensure smooth and credible elections. By 2012 elections were free, fair, and extremely competitive. They were also a purely domestic matter. Although international policy makers expressed concerns about Afghan electoral processes, they never prioritized ensuring credible elections. Fraud was perpetrated on an industrial scale in every election after 2004. Each time international officials expressed serious reservations only after the event, when it was too late to take any necessary action. At Karzai's insistence, parliamentary elections used the single non-transferable vote (SNTV) system. SNTV strongly disincentivized inclusive, broad-based parties and sharply limited elected officials' accountability to voters. Even beyond the challenges generated by SNTV, Karzai's administration consistently made the legal requirements more strenuous in order to suppress political opposition. By mid-2013 there were no legal parties at all. And those that persevered were forced to operate "informally" (U.S.

State Department 2015a: 30). Nonviolent, organized political opposition became fully dependent on remaining in the regime's good graces to survive. International proclamations about the importance of the establishing a democratic state bound by the rule of law quickly fall on deaf ears when those same actors consistently turn a blind eye toward efforts to suppress political pluralism and a rule-based legal order.

Rethinking Aid Domestically and Internationally

Rule-of-law assistance was not perfect; foreign assistance rarely is. Thus, realistic expectations are essential, because post-conflict justice systems invariably face serious challenges. But in Timor-Leste international assistance got the fundamentals right by promoting a constructive relationship between state and non-state justice as well as a more just state legal system. Aid initially focused on rebuilding a justice system that was in shambles. Not only had Indonesian officials systematically excluded the Timorese people from staffing the justice system, but the country's physical infrastructure and resources had been devastated during the referendum and its aftermath. International assistance played a substantial role in establishing and sustaining a viable state justice sector.

Advancing democracy and the rule of law demands more than just ensuring that programs meet their stated deliverables—for example, holding the required number of training sessions. Deliverables must actually result in progress toward the achievement of overarching goals. Equally important, domestic and international policy makers must be aware that even programs with ostensibly good intentions can be unhelpful or counterproductive. In Timor-Leste, Portugal explicitly and purposefully conditioned its assistance on the development of a Lusophone legal system. This produced mixed results. Portuguese aid helped establish a Lusophone state legal order that enjoyed buy-in by high-level domestic elites. At the same time, the legal system remained structurally closed to the vast majority of Timorese who did not speak Portuguese. Thus, the ideological commitment to a Lusophone legal system meant that Portuguese aid did not reach its full potential to promote access to justice and equality before the law. The inaccessibility of Portuguese and the relative disuse of Tetum within the legal system, which almost the entire population speaks, remains a serious issue that will need to be addressed before the rule of law can be fully consolidated.

Rule-of-law assistance to Afghanistan was far more problematic. The international community remained almost willfully blind as it spent immense sums without any plan for how more funding would not simply replicate previous failures (SIGAR 2015b). International actors unhesitatingly worked with institutions known to be highly corrupt and frequently enabled their corruption, which has facilitated the creation of a predatory rentier state, with the judiciary as one of its most corrupt appendages. As a result, support for the rule of law in Afghanistan was problematic both in how it was implemented and in how it was structured bureaucratically. With regard to implementation, international efforts reflected fundamental failures of timing, coordination, and strategy. Assistance during the crucial early interval and subsequently was focused mainly on unneeded and unwanted legislative reforms. Once funding began to increase dramatically, there was no plausible strategy for how it would actually advance the rule of law. Worse, once it became clear that subsidization all too often facilitated state corruption rather combated it, aid was not reconsidered, reallocated, or halted. Subsidization efforts continued and expanded with no realistic vision of how to improve assistance.

Even after these programs failed, U.S. officials showed little interest in changing their approach to dealing with institutions decidedly uninterested in the rule of law. The largest rule-of-law program, the Justice Sector Support Program, funded by the State Department, continued to focus on the same areas and activities until the fall of the regime. The USAID successor rule-of-law program, Assistance for the Development of Afghan Legal Access and Transparency, operated from 2016 to 2021. It sought to "combat[] corruption by empowering relevant Afghan Government agencies/institutions" and focused on nearly identical goals as earlier initiatives (USAID 2015: 9). The irony is presumably unintentional. Both the State Department and USAID kept doing the same things that they had always done even though there was no evidence that those programs, whether directed at state or non-state justice or a combination thereof, had ever had a positive effect on promoting the rule of law, and even though there were serious, well-documented concerns about the programs from both inside and outside the government. Indeed, in the post-2014 landscape, corruption became so deeply embedded that even Afghanistan's flagship U.S.-funded anticorruption program was found to be deeply corrupt (Lawrence 2018). Although implementers receive the most scrutiny, this perverse incentive structure is fundamentally a policy-making issue. Most nongovernmental organizations must respond to donor-established incentives to survive, and "U.S. officials chose to pursue a vision for Afghanistan's justice system that reflected American values and preferences, without sufficient regard for what was practical or possible" (SIGAR 2021a: 75). Even when the United States and other international actors belatedly recognized the importance of non-state justice, they did so almost exclusively on their terms. This meant that aid favored those most willing to work with international actors, rather than those who were the most respected or influential. In practice, meaningfully engaging non-state tribal and religious authorities in Afghanistan demands at least a degree of acceptance for sharia and tribal law codes. From a donor perspective, this can raise serious political pitfalls, since these legal systems generally fall short of international human-rights norms.

Rule-of-law programs in Timor-Leste in general made modest but positive contributions. State leaders generally supported international efforts. Program coordination was far from seamless, but there was little direct overlap, and contradictory initiatives were rare. This helped ensure that programs performed adequately and provided clearer lines of accountability. In contrast, program coordination in Afghanistan remained consistently abysmal from 2004 to 2014. Even before rule-of-law expenditures dramatically increased during Obama's Afghan surge, rule-of-law findings were opaque and largely uncoordinated, which led to a constant risk that programs would duplicate or even conflict with one another (U.S. State Department 2008: 23). Massive funding increases exacerbated these long-standing issues. It was unclear even how many programs there were or how much money had been spent (SIGAR 2015b). Throughout the period of U.S. engagement examined in this book, program strategy, implementation, and coordination problems were always being addressed, but they were never solved. Right up until the moment the Taliban retook the country, U.S. rule-of-law programs continued to be plagued by coordination problems and serious monitoring and evaluation issues (Management Systems International 2020; SIGAR 2021b).

These programing issues reflect larger, more profound problems within the U.S. foreign assistance bureaucracy. U.S. donors, particularly USAID, demand extensive control over the scope, content, and implementation of projects they support. This level of micromanagement was an especially bad fit for Afghanistan, given the fluidity of the situation and the need for a nuanced, localized approach. If programs are supposed to be dynamic, there needs to be flexibility and an understanding that calculated risks are sometimes necessary. There also needs to be a willingness to halt program activities that are not working or end partnerships with interlocutors who lack meaningful interest in promoting the rule of law. Currently, programs that fail to deliver are viewed negatively by donors and endanger future contracts. Program implementers all too often have strong incentives to collaborate with deeply corrupt state institutions and justify underperforming initiatives in their quest to secure continued funding (Bartz, Momand, and Swenson 2018).

Operational structures in Afghanistan meant that donors, and particularly the United States, could design programs and continually display almost deliberated obtuseness when it came to political realities in the recipient country. The contracting system has significant flaws. Nevertheless, by definition the contractors' job is to follow USAID's directives. More generally, this dynamic institutionalizes an unaccountability loop whereby donors can blame contractors when programs go awry, while contractors can claim that they were simply following donor instructions. Although admittedly a somewhat remote prospect, fixing this dynamic will require that Congress adopt a more nuanced understanding of how development assistance works and under what circumstances it can be most effective.

Aid can serve a variety of policy goals: some noble, some decidedly less so. Judicial state-builders must recognize the inevitability of policy trade-offs between goals. The illusion needs to be abandoned that promoting the rule of law and fostering a more constructive relationship with non-state justice can be done successfully while simultaneously accommodating a predatory regime, downplaying widespread electoral fraud, or refusing to take the core ideological foundations of non-state justice seriously. The rule of law cannot simply be tacked onto other international priorities as an afterthought. In Timor-Leste, it was a genuine priority. In Afghanistan, the international community consistently touted its commitment to strengthening the rule of law, consolidating democracy, working constructively with shuras and jirgas, and improving the quality of governance. In reality, the international community focused on stability and security, which translated into largely uncritical support for the Afghan regime. Any criticism that did occur was largely performative and rarely led to a reduction in support.

Engaging Justice Holistically

After conflict, people experience justice from both state and non-state providers. Of the two, non-state actors tend to be the most common form of justice providers. This was certainly the case in both Afghanistan and Timor-Leste. Judicial state-building in settings with a high degree of legal pluralism must make a good-faith effort to work constructively with existing pillars of legitimacy, including the non-state justice sector. Success is context-specific and would likely bear little resemblance to the justice sectors of donor states. Afghan history strongly suggests that for a state-backed justice system to project authority and possess legitimacy over wide swaths of the country, it needs to tap into religious and tribal sources of legitimacy, improve the quality of justice, and create partnerships with tribal and religious non-state justice actors. The Islamic Republic of Afghanistan failed on all three counts. But this is exactly where the rival Taliban justice system succeeded. It was essential to their campaign to return to power in Afghanistan. At least in the short to medium term, a legitimate and effective state justice system would most likely share more with the ideological foundations of the Taliban justice systems than with judicial systems found in consolidated liberal democracies.

This tension between local and international norms and ideals is inherent. This is true even in Timor-Leste, where the state enjoyed genuine legitimacy and popular support for building a modern democratic state underpinned by the rule of law. Regulations of the non-state justice sector enjoyed broad-based support and helped maintain autonomy for local communities while limiting the potential for abuses of authority. Political elites recognized the importance of non-state actors and worked to build a collaborative relationship with them. The constitution explicitly acknowledged non-state justice and its underlying values (RDTL 2002). State officials subsequently developed a sensible, minimalist framework that granted discretion to non-state authorities over small matters but funneled substantial crimes to the state courts (RDTL 2009a). Drawing on a shared vision of a state committed to development and the rule of law, as well as respect for the suco councils as an important non-state institution, the introduction and subsequent institutionalization of local elections for suco chiefs further reinforced a cooperative relationship between state and non-state actors.

International support made a modest but positive contribution here, especially after the 2006 crisis. Aid helped facilitate the transition to cooperative legal pluralism by offering it to improve the performance of non-state justice and building sensible links between local suco councils and state courts.

International engagement is notable as, by and large, it sought to help suco councils operate in areas where there was a shared commitment to improving performance and capacity as well as to following the law. Nevertheless, some tensions remained. Despite serious, sustained efforts for both the national government and the international community, issues related to the treatment of women, especially in rural and economically disadvantaged areas, remain contentious (Niner et al. 2022). Cases involving domestic violence are still frequently resolved in non-state forums despite the unequivocal status of domestic violence as a crime against the public. While yet to be fully realized, the existence of cooperative legal pluralism helped to facilitate progress toward the rule of law.

Afghanistan stands in stark contrast to Timor-Leste. While all reasonably successful historical Afghan legal orders relied heavily on religious and tribal non-state judicial authorities, the regime established after 2001 did not. The relationship with tribal authorities was competitive, and state officials did little to improve it. Eventually, in 2009, state authorities seriously considered legislative regulation of non-state justice, but the law was ill-suited to local realities and could not gain consensus within the government, let alone society more generally. A non-state justice law once again became a priority under the Ghani administration. By 2018 there was a draft on reconciliation, but again nothing was formalized (Rahbari 2018). In the end, no law was ever passed. Mere weeks before the regime collapsed, the MOJ was still holding conferences and discussing how best to approach jirgas and shuras (Ministry of Justice 2021).

The Islamic Republic of Afghanistan relied on warlords to maintain order at the local and regional levels—a fundamentally counterproductive approach to producing a just legal order. The state's inability to work constructively with religious and tribal authorities to improve the overarching law-and-order situation helped spur the Taliban's resurgence. In contrast to the state, the Taliban made establishing a legitimate legal order the core of its political program by offering inexpensive, expedient, and relatively fair dispute resolution. While locked into a combative relationship with the state legal system, the Taliban proactively built relationships with tribal and religious leaders.

International policy makers belatedly recognized non-state justice as a major pillar of Afghanistan's legal order, but they never seriously engaged with non-state justice as a preexisting structure or an independent source of legitimate authority. Instead, preferred local interlocutors were labeled the key non-state justice actors, independent of their actual social standing. These individuals were then propped up by military force and outside funds. As Thomas Johnson and Chris Mason observe, successful policy engagement with the non-state justice system of Pashtunwali demands, at a minimum, "an understanding of the core principles of this cultural value system" (Johnson and Mason 2008: 59). Although rule-of-law programs researched local areas, outputs were fixed and unrelated to local realities because counterinsurgency was always the paramount goal.

The results in Afghanistan also seriously challenged the dominant U.S. counterinsurgency view that "establishing the rule of law is a key goal and end state" (Amos and Petraeus 2008: 360). Although the rule of law is certainly worthy as both, establishing it takes far more time than a standard donor-program time frame. Before the rule of law can exist, there must be a state justice system that is reasonably legitimate and effective. The rule of law must be rooted in domestic politics and institutions. Thus, it is a long-term goal. The rule of law cannot be promoted successfully absent powerful domestic constituencies and at least some high-level state officials who take the idea seriously. International rule-of-law efforts, whether geared toward counterinsurgency or simply toward improving the quality and effectiveness of state courts, have a limited scope absent a state commitment to be bound by the law. Aid can easily become another source of income for a rentier state, warlords, and even insurgents.

While it is often cast a technocratic phenomenon, the rule of law is inherently political. Law matters, but politics matters just as much and often more. Law can be the site of fierce debates and contestations, but it lends itself to codification. By nature, politics is unpredictable, especially in post-conflict states. The creation of a state underpinned by the rule of law cannot be guaranteed even under favorable circumstances—and the circumstances that exist in post-conflict states are often far from favorable. It certainly helps when domestic and international policy makers share a commitment to the rule of law, but that cannot be taken for granted. Institutional arrangements that favor the rule of law are vital to help promote the development of the rule of law as well as to prevent those in power from using that authority to undermine limits on their authority. While there are no perfect institutions, some are better than others, and it is vital that that those institutions that are established provide positive incentives for the development of the rule of law.

The Rule of Law Is Not Built in a Day... Or a Year

The rule of law demands a legitimate state legal system, judicial independence, and a professional legal community that seeks to uphold the law. Leaders and citizens alike must abide by the law or face the consequences. The rule of law must be rooted in an executive branch that does not operate with impunity, in political parties that represent diverse societal interests and can compete for political power peacefully, and in an electoral system that encourages representation, accountability, and constructive compromise. In short, it requires a cultural commitment along with a political and institutional one. A society bound by the rule of law can be appealing to non-state authorities, as it helps to ensure that they will receive fairer treatment from the state and to maintain order on larger scale. A culture conductive to the rule of law makes citizens more likely to favor state adjudication and less likely to try to undermine the state.

Unquestionably, this is a tall order. Developing the rule of law takes time, particularly in conflict-prone settings with vibrant legal pluralism. Even in places such as Timor-Leste, which saw excellent long-term progress toward the rule of law, the day-to-day progress often seems underwhelming. Post-conflict legal orders will take many years or even decades to look like those found in most liberal democracies. Timor-Leste still faces plenty of challenges, but work there highlights the ways in which notable progress can still be achieved through thoughtful and pragmatic initiatives. Even successful programs will have modest results. These results may not lend themselves to easy quantification, and what *is* easily quantifiable may not actually matter much. It is important not to assume that just because a state (or non-state) justice sector still has significant shortcomings, programming is a failure. At the same time, as the experience in Afghanistan highlights, just because a program meets most or even all of its performance benchmarks does not mean it is actually advancing the rule of law.

Judicial state-building does not occur on a blank slate. Advancing the rule of law after conflict requires both a sensible approach toward strengthening the state justice system and realistic expectations. It also relies on constructive engagement with non-state justice actors. There is no one-size-fits-all strategy. Though the overarching legal-pluralism archetype dramatically influences which approaches are feasible, making progress in judicial state-building requires an engagement strategy that is nuanced, culturally intelligible, and credible for the particular setting. Furthermore, there are distinct strategies that domestic and international state builders can use—each with advantages and drawbacks. For instance, the decision to outsource security, and by extension local dispute resolution, to warlords allowed the international community to limit the scope of its initial engagement. However, it increased popular discontent with the state and helped give rise to an environment characterized by combative legal pluralism between the state and the Taliban.

Policy makers need a realistic judicial state-building vision that recognizes the role of non-state actors as potential allies or spoilers. They must accurately determine the overarching legal-pluralism archetype and its programmatic implications. Ideally, they should also have a credible strategy for transforming the current environment into a more constructive one. That strategy must be rooted in a deep knowledge of a country's culture, politics, and history along with a keen understanding of the potential foundations for a legitimate legal order. State builders in Timor-Leste did exactly that. Afghanistan took a very different approach. The idea that post-conflict Afghanistan would soon establish a secular legal order that wholeheartedly endorsed gender equality and international human-rights norms was optimistic to the point of absurdity. Moreover, the international community's posturing that these goals were both feasible and illustrative of the new Afghan institutions bolstered the Taliban's efforts to portray the new state as a foreign imposition that did not reflect long-standing Afghan values.

Summary

Successfully establishing the rule of law and democracy in legally pluralist settings takes generations, and the process is highly contingent. While Timor-Leste has achieved admirable progress, it has yet to consolidate either democracy or the rule of law. Afghanistan certainly faces a challenging future. The regime established in the wake of the September 11, 2001 attacks has been swept away. The Taliban's return to power was based on a successful insurgency campaign that made dispensing justice and ending corruption its focal points. Taliban justice, for all its flaws, outperformed and in a very real sense defeated its state rival. In both outcomes, domestic decisions were paramount, but the international community still offered incentives (or disincentives) that influenced both state and non-state actors.

The case-study chapters demonstrate that there were feasible, though admittedly challenging, paths toward consolidating democracy and the rule of law in both Timor-Leste and Afghanistan. These paths reflect the distinctive histories, institutions, cultures, and non-state justice systems of each country. The Timor-Leste case study demonstrates that the trajectory of state-building and the shape of the nascent state's institutions were *not* foreign impositions disconnected from society. The development of a democratic state with a legitimate legal order reflected an organic synthesis of respect for long-standing cultural values, as embodied in the suco council, with a modern, professionalized judicial system. It reflected the vision of independence held not only by political leaders but also by ordinary people. To be sure, the situation is not perfect. Corruption, for instance, remains a real problem. Nevertheless, Timor-Leste has made admirably progress toward establishing the rule of law.

The Afghanistan case study shows that the state-building project was not doomed from the outset. The new regime and its international supporters had a moment of real promise to build a state with a legitimate legal order that combined rule-of-law ideals with long-standing legal traditions. While it is impossible to predict alternative histories with certainty, in Timor-Leste and Afghanistan institutional, political, and legal choices clearly had an immense impact—and different choices would have likely led to significantly different outcomes.

Areas for Further Research

This book illustrates that the rule of law needs to be viewed holistically. It certainly includes the state legal system as commonly understood—laws, courts, judges, prosecutors, attorneys, and a whole host of related legal institutions. Beyond that, it also involves democratic governance and the institutions that underpin it. These concerns have been well documented by academics and policy makers alike. This book goes further and argues that engagement with non-state justice is a vital and understudied element in determining the success or failure of both domestic and international judicial state-building efforts. Moreover, the various forms of legal pluralism and their consequences need to be better understood. To that end, this book has explored different manifestations of legal pluralism and the strategies domestic and international actors can use to engage it in theory and in practice. It has sought to provide empirically grounded illustrations of how post-conflict states and their international backers have effectively addressed (or failed to address) the challenges of legal pluralism.

These concerns, while important, also highlight areas for additional research. First, although this study has focused on judicial state-building in settings with sustained, large-scale international state-building operations, many judicial state-building efforts in situations characterized by robust legal pluralism lack long-term, sustained international troop involvement. The presence of international peacekeepers not only influences the domestic security situation but also has major economic and political ramifications. Second, it would be worthwhile to make more comparisons across legally pluralist environments to better understand the consequences of different manifestations of legal pluralism, especially cooperative, competitive, and combative forms that enjoy substantial autonomy from the state. Third, because the overarching legal-pluralism archetype is so vital to the judicial state-building endeavor, it would be valuable to explore at a more granular level exactly how transitions to cooperative legal pluralism can best be facilitated and how to prevent competitive legal pluralism from exploding into combative legal pluralism. Finally, it would be useful to probe more deeply into how state builders can most effectively influence the behavior of domestic political elites to promote a rule-of-law culture, particularly when those elites lack a preexisting commitment to the rule of law.

Conclusion

Understanding legal pluralism is important for any legal or policy intervention, including—but by no means limited to—state-building where multiple justice systems with substantial autonomy exist. Without knowledge about the dynamics of legal pluralism, initiatives and interventions are likely to be suboptimal, ineffective, or even counterproductive. Even initiatives that enjoy short-term success are unlikely to be sustainable, for they reflect good fortune rather than an informed approach. Sound strategy requires understanding how state and non-state actors interact systematically. In other words, whether the legal pluralism environment is combative, competitive, complementary, or cooperative makes a huge difference. Domestic and international actors can use a number of potential strategies to try to influence the relationship between state and non-state justice systems, namely bridging, harmonization, incorporation, subsidization, and repression. By better understanding the archetype within which policy operates, an appropriate strategy or package of strategies for engaging with non-state actors can be selected. This by no means guarantees success, but it certainly improves the odds.

Collectively, the progress of Timor-Leste in advancing peace alongside a democratic state committed to the rule of law and the corresponding failure in Afghanistan demonstrate two key propositions. First, legal pluralism, and how domestic and international actors respond to it, really matters. These cases illuminate the importance of strategy selection by showing how policy choices can influence outcomes and that failure is rarely inevitable. Second, they show that strategy selection, while central to long-term success, is largely irrelevant if the policy intervention in question is not culturally intelligible and ultimately persuasive. Thus, the structure and implications of legal pluralism must be considered when creating and implementing policy.

Post-conflict settings are rife with both opportunity and danger. No future is predetermined. Politics is fluid by nature and invariably more so in post-conflict settings with weak institutions. Rapid changes have continued in both Timor-Leste and Afghanistan. In Timor-Leste, incumbent Prime Minister Gusmão resigned in February 2015. With Gusmano's backing, CNRT entered into a coalition with Fretilin, and Rui Maria de Araújo became prime minister (Kammen 2015). In the 2017 presidential election the Fretilin candidate, Francisco Guterres (commonly known as Lú-Olo), won with support from the coalition parties. This realignment ameliorated the partisan divisions that so threatened the country in 2006. However, it also co-opted the major opposition party and, ironically given Timor-Leste's history, consensus threatened to eliminate the partisan political competition that is so crucial for democratization, political accountability, judicial independence, and the rule of law. Consensus proved to be more of an aberration than the norm. The 2017 parliamentary elections saw a return to more traditional partisan politics, with Fretilin winning the most seats but falling short of a majority. Fretilin's Alkatari attempted to govern as a minority government, but this proved untenable. New elections were called in 2018. Fretilin won the same number of seats, but a new Gusmão-led coalition of three parties won enough seats to form a majority. Since then the president and the prime minister have repeatedly clashed—but those disagreements remained within the confines of democratic politics.

At the local level, non-state justice continues to be vibrant. In 2016, a new "Law of Sucos" aimed to coordinate suco activities with planned local governance reforms (RDTL 2016). Successful suco elections were also held that year. The culture of cooperative legal pluralism continues to deepen: suco councils still play a vibrant role in dispute resolution but also support the overarching state-building endeavor in more and more ways, ranging from long-standing concerns about ensuring communal harmony, peace, and stability to issues such as promoting good hygiene, food safety, environmental protection, and even community sporting events (RDTL 2016: Art. 6). In short, while occasionally messy, politics in Timor-Leste remain vibrantly democratic, and justice remains pluralist but cooperative.

In the midst of an active conflict, Afghanistan underwent a major transition with the election of President Ashraf Ghani in 2014. It was Afghanistan's first peaceful transfer of power. This achievement was marred by immense election fraud conducted on Ghani's behalf by outgoing President Karzai, which rendered the polls meaningless as an expression of the popular will. Ghani promised meaningful political, legal, and electoral reforms, but these either failed to materialize or delivered little. Parliamentary elections were originally slated for October 2016 but were not held until October 2018. They used the same ill-advised electoral system of SNTV. But the overall integrity of the electoral system was even worse and the environment even less secure. Owing to the Taliban insurgency, the late 2020 parliamentary elections scheduled in the province of Ghazni were never held at all.

The 2019 presidential election was even more dire. President Ghani once again faced Abdullah Abdullah in the presidential election in September 2019, and the polls were once again deeply compromised by fraud, violence, and insecurity. Ghani was eventually declared the winner in February 2020 with 50.64 percent of the vote, but these results inspired little confidence in Ghani or the political system more broadly. Abdullah once again contested the results. Voter turnout fell to record lows. In the end, only approximately 1.82 million ballots were counted, out of a pool of 9.6 million registered voters and a total population of roughly 37 million people (BBC News 2020). At the same time, the Taliban's territorial control continued to grow. Even before President Biden's announcement of unilateral U.S. military withdrawal in April 2021, the Taliban enjoyed effective control over half the country and exerted influence far beyond that (Smith 2020: 4).

On the eve of the Taliban's takeover, the state judicial system was still seen as the most corrupt part of one of the world's most corrupt states (McDevitt and Adib 2021). While justice remained the crux of the Taliban political program, as they gained even greater influence in the post-2014 period their shadow governance expanded to include areas such as education and health (Jackson and Amiri 2019). Even as shuras and jirgas remained important, Taliban justice become ever more entrenched and assertive since 2014. It presented a profound challenge to the authority of state courts. Taliban courts were able not only to "delegitimise the state and erode state justice provision," but also to "disempower and replace customary dispute resolution" in many places (Jackson and Weigand 2020: 1).

While the Taliban insurgency thrived, the authority of the Afghan state withered. Like the former Soviet-backed regime, the post-Bonn Afghan state increasingly relied on militias and warlords, with all the massive problems that dependence creates (Mashal et al. 2015). Even judged by these standards, the Islamic Republic of Afghanistan failed. President Najibullah endured after Soviet troops withdrew in 1989. After the Soviet Union itself collapsed in December 1991, it ended all external support. Najibullah's regime did not fall until April 1992. The regime of President Ghani enjoyed the promise of large-scale international support for the foreseeable future. It collapsed before international troops even left the country.

Legal pluralism in Timor-Leste and Afghanistan ultimately offers reasons for hope and trepidation. Legal pluralism can help form a vital foundation for state legitimacy. Because non-state justice is rooted in local context, understanding the different legal-pluralism archetypes and the strategies can be a helpful starting point. While the relationship between state and non-state sectors is inherently fluid, the institutions and initiatives present can help or hinder the development of more constructive relationships between state and non-state actors. Competitive legal pluralism can become cooperative. Relationships are not static; those between state and non-state justice authorities can sour as competitive relationships deteriorate into combative ones. Even situations of combative legal pluralism that seem dire are not preordained to remain that way forever. Insights from Afghanistan and Timor-Leste into judicial state-building within a competitive legal pluralist archetype offer lessons for future endeavors. And as legal pluralism has major implications for institutional design and policy initiatives in areas such as governance and development, it is not just a priority in the wake of conflict. Future research into state and non-state relationships in other settings, post-conflict and otherwise, would be a valuable next step in enabling better policy decisions through enhanced knowledge. Successfully promoting a post-conflict democratic state bound by the rule of law demands far more than understanding and addressing the challenges of legal pluralism, but it is nearly impossible without doing so.

Bibliography

Acemoglu, D., and J. Robinson (2012). *Why Nations Fail: The Origins of Power, Prosperity and Poverty*. London, Profile.

Acemoglu, D., and J. A. Robinson (2005). *Economic Origins of Dictatorship and Democracy*. Cambridge, Cambridge University Press.

Afghan International Rule of Law Programme Staffer (2014). Interview with Afghan International Rule of Law Kabul.

The Afghanistan Compact (2006). London, London Conference on Afghanistan. https://peacemaker.un.org/node/1799.

Afghanistan Research and Programme Staffer. Evaluation Unit (2015). *The A to Z Guide to Assistance in Afghanistan*. Kabul, Afghanistan Research and Evaluation Unit.

Ahmed, F. (2005). "Judicial Reform in Afghanistan: A Case Study in the New Criminal Procedure Code." *Hastings International & Comparative Law Review* **29**: 93.

Ahram, A., and C. King (2012). "The Warlord as Arbitrageur." *Theory and Society* **41**(2): 169–186.

Al Naber, M., & Molle, F. (2016). "The Politics of Accessing Desert Land in Jordan." *Land Use Policy* **59**(31): 492–503.

Albertus, M., & Menaldo, V. (2012). "Dictators as Founding Fathers? The Role of Constitutions under Autocracy." *Economics & Politics* **24**(3): 279–306.

Alexander, L., Ed. (2001). *Constitutionalism: Philosophical Foundations*. Cambridge, Cambridge University Press.

Almukhtar, S., and K. Yourish (2015). "14 Years after U.S. Invasion, the Taliban Are Back in Control of Large Parts of Afghanistan." *New York Times*. September 29. Retrieved October 26, 2015, from http://www.nytimes.com/interactive/2015/09/29/world/asia/afghanistan-taliban-maps.html?hp&action=click&pgtype=Homepage&module=photo-spot-region®ion=top-news&WT.nav=top-news&_r=0.

Amnesty International (1985). East Timor: Violations of Human Rights Extrajudicial Executions, "Disappearances," Torture and Political Imprisonment, 1975–1984. London, Amnesty International.

Amnesty International (1992). Indonesia/East Timor: "In Accordance with the Law": Statement before the United Nations Special Committee on Decolonization. London, Amnesty International.

Amnesty International (1993a). East Timor: The Unfair Political Trial of Xanana Gusmão. London, Amnesty International.

Amnesty International (1993b). Indonesia/East Timor: Seven East Timorese Still in Danger. London, Amnesty International.

Amnesty International (1995). Indonesia and East Timor: Political Prisoners and the "Rule of Law." London, Amnesty International.

Amnesty International (2011). Timor-Leste: Justice Delayed, Justice Denied: Amnesty International Submission to the UN Universal Period Review, October 2011. London, Amnesty International.

Amos, J. F., and D. H. Petraeus (2008). *The US Army/Marine Corps Counterinsurgency Field Manual 3-24*. Chicago, University of Chicago Press.

Anderson, B. (1993). "Imagining East Timor." *Arena Magazine* 4: 1–5.

Anderson, B. (2006). *Imagined Communities: Reflections on the Origin and Spread of Nationalism* (new ed). London, Verso.

Annan, K. (1999a). Progress Report of the Secretary-General on the Question of East Timor. New York, United Nations.

Annan, K. (1999b). Report of the Secretary General on the Situation in East Timor. UN Doc S/1999/1024. New York, United Nations.

Annan, K. (2004). The Rule of Law and Transitional Justice in Conflict and Post-Conflict Societies. UN doc. S/2004/616. New York, United Nations.

Annan, K. (2005). End of Mandate Report of the Secretary-General on the United Nations Mission of Support in East Timor. New York, United Nations.

Ariana News (2013). None of Political Parties Meet Legal Requirements: MOJ. Kabul, Ariana News.

Arjona, A., et al., Eds. (2015). *Rebel Governance in Civil War*. Cambridge, Cambridge University Press.

Armytage, L. (2007). "Justice in Afghanistan: Rebuilding Judicial Competence after the Generation of War." *Heidelberg Journal of International Law* 67: 185–210.

Asia Foundation (2003). Access to Justice and Legislative Development: Semi-Annual Report April 2003–September 2003. Dili, Asia Foundation.

Asia Foundation (2004). Law and Justice in East Timor: Survey of Citizen Awareness and Attitudes Regarding Law and Justice in Timor-Leste. Dili, Asia Foundation.

Asia Foundation (2005). Access to Justice and Legislative Development: Semi-Annual Report October 2004–March 2005. Dili, Asia Foundation.

Asia Foundation (2009). Access to Justice and Legislative Development: Semi-Annual Report 1 October 2008 to 31 March 2009. Dili, Asia Foundation.

Asia Foundation (2012). Access to Justice Program: End of Program Report. Dili, Asia Foundation.

AusAID (2012). Independent Evaluation of the East Timor Justice Sector Support Facility: Management Response. AusAID. Dili, AusAID.

Babo Soares, D. (2003). "Challenges for the Future." In *Out of the Ashes: Destruction and Reconstruction of East Timor*, ed. J. J. Fox and D. Babo Soares. Canberra, Australian National University E Press: 262–276.

Babo-Soares, D. (2004). "Nahe Biti: The Philosophy and Process of Grassroots Reconciliation (and Justice) in East Timor." *Asia Pacific Journal of Anthropology* 5(1): 15–33.

Baczko, A. (2016). "Legal Rule and Tribal Politics: The US Army and the Taliban in Afghanistan (2001–13)." *Development and Change* 47(6): 1412–1433.

Bain, W. (2006). "In Praise of Folly: International Administration and the Corruption of Humanity." *International Affairs* 82(3): 525–538.

Baker, B. (2010a). "Linking State and Non-State Security and Justice." *Development Policy Review* 28(5): 597–616.

Baker, B. (2010b). *Security in Post-Conflict Africa: The Role of Nonstate Policing*. Boca Raton, CRC Press.

Baker, B., and E. Scheye (2007). "Multi-Layered Justice and Security Delivery in Post-Conflict and Fragile States." *Conflict, Security & Development* 7(4): 503–528.

Balint, J., J. Evans, N. McMillan, and M. McMillan (2020). *Keeping Hold of Justice: Encounters between Law and Colonialism*. Ann Arbor, University of Michigan Press.

Barfield, T. (2008). "Culture and Custom in Nation-Building: Law in Afghanistan." *Maine Law Review* **60**: 347.

Barfield, T. (2010). *Afghanistan: A Cultural and Political History*. Princeton, Princeton University Press.

Barfield, T., et al. (2006). *The Clash of Two Goods: State and Non-State Dispute Resolution in Afghanistan*. Washington, DC, United States Institute of Peace.

Barkey, K. (2013). "Aspects of Legal Pluralism in the Ottoman Empire." In *Legal Pluralism and Empires, 1500–1850*, ed. L. Benton and R. J. Ross. New York, New York University Press: 83–107.

Barnett, M., and C. Zürcher (2009). "The Peacebuilders' Contract: How External Statebuilding Reinforces Weak Statehood." In *The Dilemmas of Statebuilding: Confronting the Contradictions of Postwar Peace Operations*, ed. R. Paris and T. D. Sisk. London, Routledge: 23–52.

Bartz, E., K. Momand., and G. Swenson (2018). "Correspondence: Debating the rule of law in Afghanistan." *International Security* **43**(1):181–185.

Bassiouni, M. C., and D. Rothenberg (2007). "An Assessment of Justice Sector and Rule of Law Reform in Afghanistan and the Need for a Comprehensive Plan." *Rome Conference "The Rule of Law in Afghanistan."*

BBC News (2020). "Afghanistan Presidential Election: Ashraf Ghani Re-elected." February 18. Retrieved December 15, 2020, from https://www.bbc.co.uk/news/world-asia-51547726.

Bearden, M. (2001). "Afghanistan, Graveyard of Empires." *Foreign Affairs* **80**(6): 17–30.

Beauvais, J. C. (2001). "Benevolent Despotism: A Critique of UN State-Building in East Timor." *New York University Journal of International Law & Policy* **33**: 1101–1178.

Beetham, D. (2013). *The Legitimation of Power*. London, Palgrave Macmillan.

Benjamin, C. (2008). "Legal Pluralism and Decentralization: Natural Resource Management in Mali." *World Development* **36**(11): 2255–2276.

Bennett, A. (2010). "Process Tracing and Causal Inference." In *Rethinking Social Inquiry: Diverse Tools, Shared Standards*, ed. H. E. Brady and D. Collier. Lanham, Rowman & Littlefield: 207–219.

Benton, L. (1994). "Beyond Legal Pluralism: Towards a New Approach to Law in the Informal Sector." *Social & Legal Studies* **3**(2): 223–242.

Benton, L. A. (2002). *Law and Colonial Cultures: Legal Regimes in World History, 1400–1900*. Cambridge, Cambridge University Press.

Berlie, J. A. (2000). "A Concise Legal History of East Timor." *Studies in Languages and Cultures of East Timor* **3**: 138–157.

Berman, P. S. (2012). *Global Legal Pluralism: A Jurisprudence of Law Beyond Borders*. Cambridge, Cambridge University Press.

Bermeo, N. (2016). "On Democratic Backsliding." *Journal of Democracy* **27**(1): 5–19.

Bhatia, M. (2007). "The Future of the Mujahideen: Legitimacy, Legacy and Demobilization in Post-Bonn Afghanistan." *International Peacekeeping* **14**(1): 90–107.

Bierschenk, T. (2008). "The Everyday Functioning of an African Public Service: Informalization, Privatization and Corruption in Benin's Legal System." *The Journal of Legal Pluralism and Unofficial Law* **40**(57): 101–139.

Bingham, T. (2011). *The Rule of Law*. London, Penguin.

Blair, R. A. (2019). "International Intervention and the Rule of Law after Civil War: Evidence from Liberia." *International Organization* **73**(2): 365–398.

Blum, G., and P. Heymann (2010). "Law and Policy of Targeted Killing." *Harvard National Security Journal* **1**: 145–170.

Bøås, M., and K. M. Jennings (2007). "'Failed States' and 'State Failure': Threats or Opportunities?" *Globalizations* 4(4): 475–485.

Boege, V., et al. (2008). "On Hybrid Political Orders and Emerging States: State Formation in the Context of 'Fragility.'" Berghof, Berghof Research Center for Constructive Conflict Management.

Böhnke, J. R., and C. Zürcher (2013). "Aid, Minds and Hearts: The Impact of Aid in Conflict Zones." *Conflict Management and Peace Science* 30(5): 411–432.

Bonn Agreement (2001). Agreement on Provisional Arrangements in Afghanistan Pending the Re-Establishment of Permanent Government Institutions. Bonn, United Nations.

Bowles, E., and T. Chopra (2008). "East Timor: Statebuilding Revisited." In *Building States to Build Peace*, ed. C. Call and V. Wyeth. Boulder, Lynne Rienner: 271–302.

Boyce, J. K. (2002). "Aid Conditionality as a Tool for Peacebuilding: Opportunities and Constraints." *Development and Change* 33(5): 1025–1048.

Bradley, C. M. (2012). "Criminal Law: Is the Exclusionary Rule Dead?" *Journal of Criminal Law & Criminology* 102(1): 1–23.

Braithwaite, J., et al. (2012). *Networked Governance of Freedom and Tyranny: Peace in Timor-Leste*. Canberra, ANU E Press.

Brown, F. Z. (2012). *The US Surge and Afghan Local Governance*. Washington, DC, U.S. Institute of Peace.

Brown, M. A. (2012). "Entangled Worlds: Villages and Political Community in Timor-Leste." *Local-Global: Identity, Security, Community* 11: 54–71.

Brown, M. A., and A. F. Gusmao (2009). "Peacebuilding and Political Hybridity in East Timor." *Peace Review* 21(1): 61–69.

Bush, G. W. (2002). The National Security Strategy of the United States of America. Washington, DC, Executive Office of the President.

Butler, M. J. (2012). "Ten Years After: (Re)Assessing Neo-Trusteeship and UN State-Building in Timor-Leste." *International Studies Perspectives* 13(1): 85–104.

Call, C. T., and V. Wyeth, Eds. (2008). *Building States to Build Peace*. Boulder, Lynne Rienner.

Call, C. T. (2008a). "Ending Wars, Building States." In *Building States to Build Peace*, ed. C. T. Call and V. Wyeth. Boulder, Lynne Rienner: 1–19.

Call, C. T. (2008b). "The Fallacy of the 'Failed State.'" *Third World Quarterly* 29(8): 1491–1507.

Call, C. T. (2012). *Why Peace Fails: The Causes and Prevention of Civil War Recurrence*. Washington, DC, Georgetown University Press.

Call, C. T., and E. M. Cousens (2008). "Ending Wars and Building Peace: International Responses to War-Torn Societies." *International Studies Perspectives* 9(1): 1–21.

Campbell, M., and G. Swenson (2016). "Legal Pluralism andWomen's Rights after Conflict: The Role of CEDAW." *Columbia Human Rights Law Review* 48(1): 112–146.

Canfield, R. (1986). "Ethnic, Regional, and Sectarian Alignments in Afghanistan." In *The State, Religion and Ethnic Politics: Afghanistan, Iran, Pakistan*, ed. A. Banuazizi and M. Weiner. Syracuse, Syracuse University Press: 75–103.

Cao, L. (2016). *Culture in Law and Development: Nurturing Positive Change*. New York, Oxford University Press.

Caplan, R. (2005). *International Governance of War-Torn Territories: Rule and Reconstruction*. Oxford, Oxford University Press.

Caplan, R. (2012). "Exit Strategies and State Building." In *Exit Strategies and State Building*, ed. R. Caplan. Oxford, Oxford University Press.

Cappelletti, M. (1993). "Alternative Dispute Resolution Processes within the Framework of the World-Wide Access-to-Justice Movement." *The Modern Law Review* 56(3): 282–296.

Carothers, T. (1998). "The Rule of Law Revival." *Foreign Affairs* 77(2): 95–106.

Carothers, T. (1999). *Aiding Democracy Abroad: The Learning Curve*. Washington, DC, Carnegie Endowment for International Peace.

Carothers, T. (2003). Promoting the Rule of Law Abroad: The Problem of Knowledge. Washington, DC, Carnegie Endowment for International Peace.

Carothers, T. (2006a). "The Backlash against Democracy Promotion." *Foreign Affairs* 85(2): 55–68.

Carothers, T. (2006b). *Confronting the Weakest Link: Aiding Political Parties in New Democracies*. Washington, DC, Carnegie Endowment for International Peace.

Carothers, T. (2009). Revitalizing Democracy Assistance: The Challenge of USAID. Washington, DC, Carnegie Endowment for International Peace.

Carter, L., and K. Connor (1989). *A Preliminary Investigation of Contemporary Afghan Councils*. Peshawar, Agency Coordinating Body for Afghan Relief.

Center for Policy and Human Development and UNDP (2007). Afghanistan Human Development Report 2007: Bridging Modernity and Tradition; Rule of Law and the Search for Justice. Kabul, UNDP.

Central Intelligence Agency (2015). The World Factbook. Washington, DC, Central Intelligence Agency.

Centre for International Governance Innovation (2011). Security Sector Reform Monitor: Timor-Leste. Waterloo, Centre for International Governance Innovation 4.

Chandler, D. C. (2006). *Empire in Denial: The Politics of State-Building*. London, Pluto.

Channell, W. (2006). "Lessons Not Learned about Legal Reform." In *Promoting the Rule of Law Abroad*, ed. T. Carothers. Washington, DC, Carnegie Endowment for International Peace: 137–159.

Chaudhuri, R., and T. Farrell (2011). "Campaign Disconnect: Operational Progress and Strategic Obstacles in Afghanistan, 2009–2011." *International Affairs* 87(2): 271–296.

Chayes, S. (2015). *Thieves of State: Why Corruption Threatens Global Security*. New York, W. W. Norton.

Checchi and Company Consulting (2006). Contract No. Dfd-1-00-04-00170-00: Twelfth Quarterly Performance Monitoring Report for the Period July 1 to September 30, 2006. Kabul, USAID.

Checchi and Company Consulting (2009a). Contract No. Dfd-1-00-04-00170-00: Eighteenth Quarterly Performance Monitoring Report for the Period January 1 to March 31, 2009. Kabul, USAID.

Checchi and Company Consulting (2009b). Final Report of the Afghanistan Rule of Law Project. Kabul, USAID.

Checchi and Company Consulting (2010). Rule of Law Stabilization Program-Informal Component; Monthly Report—July 2010. Kabul, USAID.

Checchi and Company Consulting (2011). Rule of Law Stabilization Program-Informal Compnent: Final Report. Kabul, USAID.

Checchi and Company Consulting (2012a). Performance Monitoring Plan 14 October 2012–13 January 2014; Rule of Law Stabilization Program–Informal Component. Kabul, USAID.

Checchi and Company Consulting (2012b). RLS-I Impact Evaluation Report, July 2012. Kabul, USAID.

Checchi and Company Consulting (2013). Performance Monitoring Plan 14 October 2012–13 January 2014 [revised version]. Kabul, USAID.
Checchi and Company Consulting (2014a). Final Evaluation Report: Rule of Law Stabilization Program—Informal Component. Kabul, USAID.
Checchi and Company Consulting (2014b). Monthly Report, January 2014 Rule of Law Stabilization Program—Informal Component; Contract Number: Aid-306-C-12-00013. Kabul, USAID.
Chesterman, S. (2002). "Walking Softly in Afghanistan: The Future of UN State-Building." *Survival* 44(3): 37–45.
Chesterman, S. (2004). *You, the People: The United Nations, Transitional Administration, and State-Building*. Oxford, Oxford University Press.
Chopra, J. (2000). "The UN's Kingdom of East Timor." *Survival* 42(3): 27–40.
Chopra, J. (2002). "Building State Failure in East Timor." *Development and Change* 33(5): 979–1000.
Chopra, T. (2008). Building Informal Justice in Northern Kenya: Justice for the Poor and Legal Resources Foundation Trust Research Report. Washington, DC, World Bank.
Chopra, T., and D. Isser (2012). "Access to Justice and Legal Pluralism in Fragile States: The Case of Women's Rights." *Hague Journal on the Rule of Law* 4(02): 337–358.
Chopra, T., et al. (2009). Fostering Justice in Timor-Leste: Rule of Law Program Evaluation. Dili, USAID.
Ciorciari, J. D. (2021). *Sovereignty Sharing in Fragile States*. Redwood City: Stanford University Press.
Clark, K. (2011). The Layha: Calling the Taleban to Account. Kabul, Afghanistan Analysts Network.
Clark, P. (2007). "Hybridity, Holism, and Traditional Justice: The Case of the Gacaca Courts in Post-Genocide Rwanda." *George Washington International Law Review* 39: 765–838.
Clark, P. (2010). *The Gacaca Courts, Post-Genocide Justice and Reconciliation in Rwanda: Justice without Lawyers*. Cambridge, Cambridge University Press.
Clark, R. S. (1980). "Decolonization of East Timor and the United Nations Norms on Self-Determination and Aggression." *Yale Journal of World Public Order* 7: 2–44.
Clark, R. S. (2000). "East Timor, Indonesia, and the International Community." *Temple International & Comparative Law Journal* 14: 75–87.
Clements, P. (2020). "Improving Learning and Accountability in Foreign Aid." *World Development* 125: 104670.
Cliffe, S., and N. Manning (2008). "Practical Approaches to Building State Institutions." In *Building States to Build Peace*, ed. C. Call and V. Wyeth. Boulder, Lynne Rienner.
Coburn, N. (2013). Informal Justice and the International Community in Afghanistan. Washington, DC, United States Institute of Peace.
Coburn, N., and A. Larson (2014). *Derailing Democracy in Afghanistan: Elections in an Unstable Political Landscape*. New York, Columbia University Press.
Coghlan, M., and S. Hayati (2012). Final Evaluation of Component One: The Access to Justice Program. Dili, USAID.
Cohen, D. (2007). "Hybrid Justice in East Timor, Sierra Leone, and Cambodia: Lessons Learned and Prospects for the Future." *Stanford Journal of International Law* 43: 1–38.
Cohen, E., et al. (2011). "Truth and Consequences in Rule of Law: Inferences, Attribution and Evaluation." *Hague Journal on the Rule of Law* 3(01): 106–129.
Collier, D. (2011). "Understanding Process Tracing." *PS: Political Science & Politics* 44(04): 823–830.

Collier, P., A. Hoeffler, and M. Söderbom (2008). "Post-Conflict Risks." *Journal of Peace Research* **45**(4): 461–478.

Collins, J. J. (1986). *The Soviet Invasion of Afghanistan: A Study in the Use of Force in Soviet Foreign Policy*. Lexington, Lexington Books.

Commission for Reception, Truth and Reception (2005). Chega! The Report of the Commission for Reception, Truth and Reconciliation in Timor-Leste. Dili, Commission for Reception, Truth, and Reconciliation.

Connolly, B. (2005). "Non-State Justice Systems and the State: Proposals for a Recognition Typology." *Connecticut Law Review* **38**: 239–294.

Cooley, A., and J. Ron (2002). "The NGO Scramble: Organizational Insecurity and the Political Economy of Transnational Action." *International Security* **27**(1): 5–39.

Coordination Team of the UN System High-Level Task Force on the Global Food Security Crisis (2010). Timor Leste: Full Country Visit Report—7 to 14 November 2009. New York, United Nations.

Corcoran-Nantes, Y. (2009). "The Politics of Culture and the Culture of Politics—a Case Study of Gender and Politics in Lospalos, Timor-Leste." *Conflict, Security & Development* **9**(2): 165–187.

Cotton, J. (2007). "Timor-Leste and the Discourse of State Failure." *Australian Journal of International Affairs* **61**(4): 455–470.

CNRT (1998). Magna Carta Concerning Freedoms, Rights, Duties and Guarantees for the People of East Timor. Peniche, Portugal.

Crouch, J. (2008). "The Law: Presidential Misuse of the Pardon Power." *Presidential Studies Quarterly* **38**(4): 722–734.

Cummins, D. (2010). Local Governance in Timor-Leste: The Politics of Mutual Recognition. Ph.d. diss. Sydney, University of New South Wales.

Cummins, D., and M. F. P. Guterres (2012). "Ami Sei Vítima Beibeik": Looking to the Needs of Domestic Violence Victims. Dili, Asia Foundation.

Cummins, D., and V. Maia (2012). Community Experiences of Decentralised Development in Timor-Leste. Dili, Asia Foundation.

Daniels, R. J., and M. Trebilcock (2004). "The Political Economy of Rule of Law Reform in Developing Countries." *Michigan Journal of International Law* **26**: 99–140.

De Lauri, A. (2013a). "Access to Justice and Human Rights in Afghanistan." *Crime, Law and Social Change* **60**(3): 261–285.

De Lauri, A. (2013b). "Corruption, Legal Modernisation and Judicial Practice in Afghanistan." *Asian Studies Review* **37**(4): 527–545.

De Waal, A. (2015). *The Real Politics of the Horn of Africa: Money, War and the Business of Power*. Cambridge, Polity Press.

Democratic Republic of Afghanistan (1980). Fundamental Principles of the Democratic Republic of Afghanistan (Interim Constitution of Afghanistan). Kabul, Democratic Republic of Afghanistan.

Democratic Republic of Afghanistan (1987). Constitution of Afghanistan. Kabul, Democratic Republic of Afghanistan.

Department of Defense (2012a). Report on Progress toward Security and Stability in Afghanistan. Washington, DC, Department of Defense.

Department of Defense (2012b). Report on Progress toward Security and Stability in Afghanistan/ United States Plan for Sustaining the Afghan National Security Forces. Washington, DC, Department of Defense.

Department of Defense Centeral Command (2010). "Rule of Law Conference Brings Together Afghan, International Partners." Retrieved June 27, 2022, from https://www.centcom.mil/MEDIA/PRESS-RELEASES/Press-Release-View/Article/903828/rule-of-law-conference-brings-together-afghan-international-partners/.

Dezalay, Y., and B. G. Garth (1998). *Dealing in Virtue: International Commercial Arbitration and the Construction of a Transnational Legal Order*. Chicago, University of Chicago Press.

DFID (2004). Non-State Justice and Security Systems. London, DFID.

Diamond, L. (1999). *Developing Democracy: Toward Consolidation*. Baltimore, Johns Hopkins University Press.

Diamond, L. (2008). *The Spirit of Democracy: The Struggle to Build Free Societies Throughout the World*. New York, Henry Holt.

Diamond, L., and R. Gunther, Eds. (2001). *Political Parties and Democracy*. Baltimore, Johns Hopkins University Press.

Dobbins, J., et al. (2003). *America's Role in Nation-Building: From Germany to Iraq*. Santa Monica, RAND Corporation.

Dobbins, J., et al. (2007). *The Beginner's Guide to Nation-Building*. Santa Monica, RAND Corporation.

Dorman, S. R. (2006). "Post-Liberation Politics in Africa: Examining the Political Legacy of Struggle." *Third World Quarterly* 27(6): 1085–1101.

Dorronsoro, G. (2005). *Revolution Unending: Afghanistan, 1979 to the Present*. New York, Columbia University Press.

Downer, A. (2000). "East Timor—Looking Back on 1999." *Australian Journal of International Affairs* 54(1): 5–10.

Downs, G., and S. J. Stedman (2002). "Evaluation Issues in Peace Implementation." In *Ending Civil Wars: The Implementation of Peace Agreements*, ed. S. J. Stedman, D. Rothchild and E. M. Cousens. Boulder, Lynne Rienner: 43–69.

Doyle, M. W., and N. Sambanis (2006). *Making War and Building Peace: United Nations Peace Operations*. Princeton, Princeton University Press.

Dunn, D., et al. (2011). Assessment: Afghanistan Rule of Law Stabilization Program (Informal Component); Final Report. Kabul, USAID.

Dunn, J. (1983). *Timor: A People Betrayed*. Milton, Queensland, Jacaranda Press.

Dunning, T. (2011). "Fighting and Voting: Violent Conflict and Electoral Politics." *Journal of Conflict Resolution* 55(3): 327–339.

Dupree, L. (1973). *Afghanistan*. Princeton, Princeton University Press.

Edelstein, D. M. (2009). "Foreign Militaries, Sustainable Institutions, and Postwar Statebuilding." In *The Dilemmas of Statebuilding: Confronting the Contradictions of Postwar Peace Operations*, ed. R. Paris and T. D. Sisk. London, Routledge: 81–103.

Edwards, D. B. (2002). *Before Taliban: Genealogies of the Afghan Jihad*. Berkeley, University of California Press.

Edwards, H. T. (1986). "Alternative Dispute Resolution: Panacea or Anathema?" *Harvard Law Review* 99(3): 668–684.

Eikenberry, K. W. (2013). "The Limits of Counterinsurgency Doctrine in Afghanistan: The Other Side of the Coin." *Foreign Affairs* 92(5): 59–74.

Elgie, R. (2005). "Variations on a Theme." *Journal of Democracy* 16(3): 98–112.

Elkins, Z., et al. (2007). "Baghdad, Tokyo, Kabul . . . Constitution Making in Occupied States." *William and Mary Law Review* 49: 1139–1178.

Elkins, Z., et al. (2009). *The Endurance of National Constitutions*. Cambridge, Cambridge University Press.

Ellickson, R. C. (1991). *Order without Law: How Neighbors Settle Disputes*. Cambridge, Harvard University Press.

Engel, R. (2007). "The Building of Timor-Leste: International Contributions to a Fragile State." New York, Centre for International Conflict Resolution, Columbia University.

Everett, S. (2009). Law and Justice in Timor-Leste: A Survey of Citizen Awareness and Attitudes Regarding Law and Justice 2008. Dili, Asia Foundation.

Farquharson, K. (2005). "A Different Kind of Snowball: Identifying Key Policymakers." *International Journal of Social Research Methodology* **8**(4): 345–353.

Farran, S. (2006). "Is Legal Pluralism an Obstacle to Human Rights-Considerations from the South Pacific?" *Journal of Legal Pluralism & Unofficial Law* **52**: 77–105.

Faundez, J. (2011). "Legal Pluralism and International Development Agencies: State Building or Legal Reform?" *Hague Journal on the Rule of Law* **3**(01): 18–38.

Fearon, J. D., and D. D. Laitin (2003). "Ethnicity, Insurgency, and Civil War." *The American Political Science Review* **97**(1): 75–90.

Fearon, J. D., and D. D. Laitin (2004). "Neotrusteeship and the Problem of Weak States." *International Security* **28**(4): 5–43.

Fernandes, M. S. (2010). "The Ongoing Crisis in East-Timor: Analysis of Endogenous and Exogenous Factors." In *State, Society and International Relations in Asia*, ed. M. Parvizi Amineh. Amsterdam, Amsterdam University Press: 119–132.

Fishman, R. M. (2011). "Democratic Practice after the Revolution: The Case of Portugal and Beyond." *Politics & Society* **39**(2): 233–267.

Fitzpatrick, P. (1984). "Traditionalism and Traditional Law." *Journal of African Law* **28**(1–2): 20–27.

Ford, S. (2020). "Has President Trump Committed a War Crime by Pardoning War Criminals?" *American University International Law Review* **35**: 757–820.

Former Afghan Judge (2014). Interview with Former Afghan Judge. Kabul.

Forsyth, M. (2007). "Typology of Relationships between State and Non-State Justice Systems, A." *Journal of Legal Pluralism & Unofficial Law* **56**: 67–113.

Forsyth, M. (2009). *A Bird That Flies with Two Wings: The Kastom and State Justice Systems in Vanuatu*. Canberra, ANU E Press.

Fowler, A. (2013). *Striking a Balance: A Guide to Enhancing the Effectiveness of Non-Governmental Organisations in International Development*. New York, Routledge.

Franck, T. M. (2001). "Are Human Rights Universal?" *Foreign Affairs* **80**(1): 191–204.

Frantz, E., and E. A. Stein (2017). Countering Coups: Leadership Succession Rules in Dictatorships. *Comparative Political Studies* 50(7): 935–962.

Freedom House and American Bar Association Rule of Law Initiative (2007). Rule of Law in Timor-Leste. Dili, USAID.

Frigaard, S., et al. (2008). UNDP Strengthening the Justice System in Timor-Leste Programme: Independent/External Mid-Term Evaluation Report September 2007. Oslo, Norwegian Agency for Development Cooperation.

Fu, H. (2010). "Access to Justice in China." In *Legal Reforms in China and Vietnam*, ed. J. Gilliespie and A. H. Y. Chen. New York, Routledge: 164–187.

Fukuyama, F. (2004). *State-Building: Governance and World Order in the 21st Century*. Ithaca, Cornell University Press.

Fukuyama, F. (2010). "Transitions to the Rule of Law." *Journal of Democracy* **21**(1): 33–44.

Fukuyama, F. (2011). *The Origins of Political Order: From Prehuman Times to the French Revolution*. New York, Farrar, Straus & Giroux.

Fukuyama, F. (2014). *Political Order and Political Decay: From the Industrial Revolution to the Globalization of Democracy*. London, Profile.

Gaitis, J. M. (2004). "International and Domestic Arbitration Procedure: The Need for a Rule Providing a Limited Opportunity for Arbitral Reconsideration of Reasoned Awards." *American Review of International Arbitration* **15**: 9–639.

Gall, C. (2014). "In Afghan Election, Signs of Systemic Fraud Cast Doubt on Many Votes." *The New York Times*, August 23.

Galtung, J. (1969). "Violence, Peace, and Peace Research." *Journal of Peace Research* **6**(3): 167–191.

Gama, P. (1995). "The War in the Hills, 1975–85: A Fretlin Commander Remembers." In ed P. Carey and G. C. Bentley. *East Timor at the Crossroads: The Forging of a Nation*. London, Cassell: 97–105.

Gaston, E., and E. Jensen (2016). "Rule of Law and Statebuilding in Afghanistan: Testing Theory with Practice." In *State-Strengthening in Afghanistan: Lessons Learned 2001–14*, ed. S. Smith and C. Cookman. Washington, DC, United States Institute for Peace: 69–79.

George, A. L., and A. Bennett (2005). *Case Studies and Theory Development in the Social Sciences*. Cambridge, MIT Press.

Gerring, J. (2004). "What Is a Case Study and What Is It Good For?" *The American Political Science Review* **98**(2): 341–354.

Gerring, J. (2007). *Case Study Research: Principles and Practices*. Cambridge, Cambridge University Press.

Ghani, A. (1978). "Islam and State-Building in a Tribal Society Afghanistan: 1880—1901." *Modern Asian Studies* **12**(02): 269–284.

Ghani, A., and C. Lockhart (2009). *Fixing Failed States: A Framework for Rebuilding a Fractured World*. Oxford, Oxford University Press.

Giddens, A. (1993). *Sociology*. Cambridge, Polity Press.

Ginsburg, T. (2011). "An Economic Interpretation of the Pashtunwalli." *University of Chicago Legal Forum* **2011**: 89–114.

Ginsburg, T., and A. Huq (2014). "What Can Constitutions Do? The Afghan Case." *Journal of Democracy* **25**(1): 116–130.

Ginsburg, T., and T. Moustafa (2008a). "Introduction: The Functions of Courts in Authoritarian Politics." In *Rule by Law: The Politics of Courts in Authoritarian Regimes*, ed. T. Ginsburg and T. Moustafa. Cambridge, Cambridge University Press: 1–22.

Ginsburg, T., and T. Moustafa, Eds. (2008b). *Rule by Law: The Politics of Courts in Authoritarian Regimes*. Cambridge, Cambridge University Press.

Girod, D. (2015). *Explaining Post-Conflict Reconstruction*. Oxford, Oxford University Press.

Giustozzi, A. (2012a). *Empires of Mud: Wars and Warlords in Afghanistan*. New York, St. Martin's Press.

Giustozzi, A. (2012b). "Hearts, Minds, and the Barrel of a Gun: The Taliban's Shadow Government." *Prism: a Journal of the Center for Complex Operations* **3**(2): 71.

Giustozzi, A. (2013). "March towards Democracy? The Development of Political Movements in Afghanistan." *Central Asian Survey* **32**(3): 318–335.

Giustozzi, A. (2014). "The Taliban's 'Military Courts.'" *Small Wars & Insurgencies* **25**(2): 284–296.

Giustozzi, A., and A. Baczko (2014). "The Politics of the Taliban's Shadow Judiciary, 2003–2013." *Central Asian Affairs* **1**(2): 199–224.

Giustozzi, A., C. Franco, and A. Baczko (2013). *Shadow Justice: How the Taliban Run Their Judiciary*. Kabul, Integrity Watch Afghanistan.

Glatzer, B. (1998). "Is Afghanistan on the Brink of Ethnic and Tribal Disintegration?" In *Fundamentalism Reborn? Afghanistan and the Taliban*, ed. W. Maley. London, Hurst: 167–181.

Gledhill, J. (2012). "Competing for Change: Regime Transition, Intrastate Competition, and Violence." *Security Studies* **21**(1): 43–82.

Gledhill, J. (2014). "A Confluence of Competitions: Regime-Building and Violence in Timor-Leste." *Asian Security* **10**(2): 123–150.

Glenn, H. P. (2011). "Sustainable Diversity in Law." *Hague Journal on the Rule of Law* 3(01): 39–56.

Global Policy Forum (2004). "East Timor Says No to UN Tribunal." *Laksamana.Net 9*. Retrieved February 10, 2014, from https://http://www.globalpolicy.org/component/content/article/163/29195.html.

Gohari, M. J. (2000). *The Taliban: Ascent to Power*. Oxford, Oxford University Press.

Goldstone, A. (2012). "East Timor." In *Exit Strategies and State Building*, ed. R. Caplan. Oxford, Oxford University Press: 175–193.

Goodhand, J. (2008). "Corrupting or Consolidating the Peace? The Drugs Economy and Post-Conflict Peacebuilding in Afghanistan." *International Peacekeeping* **15**(3): 405–423.

Goodson, L. P. (2001). *Afghanistan's Endless War: State Failure, Regional Politics, and the Rise of the Taliban*. Seattle, University of Washington Press.

Goodson, L. P. (2005). "Afghanistan in 2004: Electoral Progress and an Opium Boom." *Asian Survey* **45**(1): 88–97.

Government of Australia (2007). East Timor—Justice Sector Support Facility: Design Framework for Goa Assistance. Canberra, Government of Australia.

Graydon, C. (2005). "Local Justice Systems in Timor-Leste: Washed Up, or Watch This Space?" *Development Bulletin, Canberra, Australian National University* **68**: 66–70.

Graydon, C. (2011). Mid Term Evaluation of Access to Justice Program. Dili, Asia Foundation.

Greentree, T. (2013). "Bureaucracy Does Its Thing: US Performance and the Institutional Dimension of Strategy in Afghanistan." *Journal of Strategic Studies* **36**(3): 325–356.

Gregorian, V. (1969). *The Emergence of Modern Afghanistan: Politics of Reform and Modernization, 1880–1946*. Stanford, Stanford University Press.

Grenfell, L. (2006). "Legal Pluralism and the Rule of Law in Timor Leste." *Leiden Journal of International Law* **19**(2): 305–337.

Grenfell, L. (2009). "Promoting the Rule of Law in Timor-Leste." *Conflict, Security & Development* **9**(2): 213–238.

Griffiths, J. (1986). "What Is Legal Pluralism?" *Journal of Legal Pluralism & Unofficial Law* **24**: 1–56.

Grimm, S., and J. Leininger (2012). "Not All Good Things Go Together: Conflicting Objectives in Democracy Promotion." *Democratization* **19**(3): 391–414.

GTZ (2014). Promoting the Rule of Law: Rights and Security for All Afghans. Kabul, GTZ.

Guevara, C. (1969). *Guerrilla Warfare*. Middlesex, Penguin Books.

Gunn, G. C. (1997). *East Timor and the United Nations: The Case for Intervention*. Lawrenceville, Red Sea Press.

Gunther, R., and L. Diamond (2001). "Introduction." In *Political Parties and Democracy*, ed. L. Diamond and R. Gunther. Baltimore, Johns Hopkins University Press.

Gunther, R., and L. Diamond (2003). "Species of Political Parties a New Typology." *Party Politics* **9**(2): 167–199.

Gusmão, X. (2009). Intervention: Response by His Excellency the Prime Minister Kay Rala Xanana Gusmao on the Occasion of the Motion of No Confidence. Dili, RDTL.

Habermas, J. (1995). "On the Internal Relation between the Rule of Law and Democracy." *European Journal of Philosophy* 3(1): 12–20.

Hager, R. (1983). "State, Tribe and Empire in Afghan Interpolity Relations." In *The Conflict of Tribe and State in Iran and Afghanistan*, ed. R. Tapper. London, Croom Helm: 83–118.

Hamidi, F., and A. Jayakody (2015). Separation of Powers under the Afghan Constitution: A Case Study. Kabul, Afghanistan Research and Evaluation Unit.

Harper, E. (2011a). Engaging with Customary Justice Systems. Rome, International Law Development Organization.

Harper, E., Ed. (2011b). *Working with Customary Justice Systems: Post-Conflict and Fragile States*. Rome, International Development Law Organization.

Harrison, L. E., and S. P. Huntington, Eds. (2000). *Culture Matters: How Values Shape Human Progress*. New York, Basic Books.

Hart, J. T. (1987). "Aspects of Criminal Justice." In *Indonesia and the Rule of Law: Twenty Years of "New Order" Government*, ed. H. Thoolen. London, Frances Pinter: 166–209.

Hartmann, M. E., and A. Klonowiecka-Milart (2011). "Lost in Translation: Legal Transplants without Consensus-Based Adaption." In *The Rule of Law in Afghanistan: Missing in Inaction*, ed. W. Mason. Cambridge, Cambridge University Press: 266–298.

Hassan, O. (2020). "The Evolution of the European Union's Failed Approach to Afghanistan." *European Security* 29(1): 74–95.

Hathaway, O. A. (2002). "Do Human Rights Treaties Make a Difference?" *The Yale Law Journal* 111(8): 1935–2042.

Havercroft, J., et al. (2018). "Editorial: Donald Trump as Global Constitutional Breaching Experiment." *Global Constitutionalism* 7(1): 1–13. Retrieved June 27, 2022 from https://www.cambridge.org/core/journals/global-constitutionalism/article/editorial-donald-trump-as-global-constitutional-breaching-experiment/3A56DF4FB35F0507AEF148A4512A330E.

Hechter, M. (2009). "Legitimacy in the Modern World." *American Behavioral Scientist* 53: 279–288.

Hehir, A. (2007). "The Myth of the Failed State and the War on Terror: A Challenge to the Conventional Wisdom." *Journal of Intervention and Statebuilding* 1(3): 307–332.

Helmke, G., and Levitsky, S. (2004). "Informal Institutions and Comparative Politics: A Research Agenda." *Perspectives on Politics* 2(4): 725–740.

Hensler, D. R. (2004). "Our Courts, Ourselves: How the Alternative Dispute Resolution Movement Is Re-Shaping Our Legal System." *Pennsylvania State Law Review* 108: 165–197.

Hilbink, L. (2007). *Judges beyond Politics in Democracy and Dictatorship: Lessons from Chile*. Cambridge, Cambridge University Press.

Hironaka, A. (2009). *Neverending Wars: The International Community, Weak States, and the Perpetuation of Civil War*. Cambridge, Harvard University Press.

Hogg, R., et al. (2013). *Afghanistan in Transition: Looking beyond 2014*. Washington, DC, World Bank.

Hohe, T. (2002a). "The Clash of Paradigms: International Administration and Local Political Legitimacy in East Timor." *Contemporary Southeast Asia* 24(3): 569–589.

Hohe, T. (2002b). "'Totem Polls': Indigenous Concepts and 'Free and Fair' Elections in East Timor." *International Peacekeeping* 9(4): 69–88.

Hohe, T. (2003). "Justice without Judiciary in East Timor." *Conflict, Security & Development* 3(3): 335–357.
Hohe, T., and R. Nixon (2003). *Reconciling Justice: "Traditional" Law and State Judiciary in East Timor*. Washington, DC, United States Institute of Peace.
Horowitz, D. L. (2000). *Ethnic Groups in Conflict*. Berkeley, University of California Press.
Horowitz, D. L. (2006). "Constitutional Courts: A Primer for Decision Makers." *Journal of Democracy* 17(4): 125–137.
Howard, L. M. (2008). *UN Peacekeeping in Civil Wars*. Cambridge, Cambridge University Press.
Huntington, S. P. (1993). *The Third Wave: Democratization in the Late Twentieth Century*. Norman, University of Oklahoma Press.
Huntington, S. P. (2006). *Political Order in Changing Societies*. New Haven, Yale University Press.
IDLO (2010). Legal and Judicial Development Assistance Global Report 2010. Rome, IDLO.
IDLO (2013). SIGAR's Letter to Secretary John Kerry: Incorrect. Kabul, IDLO.
International Crisis Group (2003). Afghanistan: The Constitutional Loya Jirga. Kabul, International Crisis Group.
International Crisis Group (2005). Political Parties in Afghanistan. Asia Policy Briefing 39. Kabul, International Crisis Group.
International Crisis Group (2008). Afghanistan: The Need for International Resolve. Kabul, International Crisis Group.
International Crisis Group (2012). Afghanistan: The Long, Hard Road to the 2014 Transition. Kabul, International Crisis Group.
International Crisis Group (2013). Afghanistan's Parties in Transition. Kabul, International Crisis Group.
International Crisis Group (2020). Taking Stock of the Taliban's Perspectives on Peace. Brussels, International Crisis Group.
International NGO Manager (2014). Interview with International NGO Manager in Timor-Leste. Dili.
International NGO Professional (2014a). Interview with International NGO Professional in Afghanistan. Kabul.
International NGO Professional (2014b). Interview with International NGO Professional in Timor-Leste. Dili.
International Rule of Law Professional (2015). Skype Interview with International Rule of Law Professional in Afghanistan.
Islamic Republic of Afghanistan (2004a). Constitution of the Islamic Republic of Afghanistan. Kabul, Islamic Republic of Afghanistan.
Islamic Republic of Afghanistan (2004b). Interim Criminal Procedure Code for Courts. Kabul, Islamic Republic of Afghanistan.
Islamic Republic of Afghanistan (2005). Law of the Organization and Authority of the Courts of Islamic Republic of Afghanistan. Kabul, Islamic Republic of Afghanistan.
Islamic Republic of Afghanistan (2008). Afghanistan National Development Strategy 1387 – 1391 (2008–2013). Kabul, Islamic Republic of Afghanistan.
Islamic Republic of Afghanistan (2009). Political Parties Law. Kabul, Islamic Republic of Afghanistan.
Islamic Republic of Afghanistan (2010). Afghanistan National Development Stratergy: Prioritization and Implementation Plan: Mid 2010–Mid 2013; Volume I. Kabul, Islamic Republic of Afghanistan.

Isser, D. (2011a). "Conclusion: Understanding and Engaging Customary Justice Systems." In *Customary Justice and the Rule of Law in War Torn Societies*, ed. D. Isser. Washington, DC, U.S. Institute of Peace Press: 325–367.

Isser, D., Ed. (2011b). *Customary Justice and the Rule of Law in War-Torn Societies*. Washington, DC, United States Institute of Peace.

Jackson, A., and R. Amiri (2019). Insurgent Bureaucracy: How the Taliban Makes Policy. Wahington, DC, United States Institute of Peace.

Jackson, A., and F. Weigand (2020). Rebel Rule of Law: Taliban Courts in the West and North-west of Afghanistan. London, Overseas Development Institute.

Jahn, B. (2007). "The Tragedy of Liberal Diplomacy: Democratization, Intervention, Statebuilding (Part II)." *Journal of Intervention and Statebuilding* 1(2): 211–229.

Jalali, A. A. (2003). "Afghanistan in 2002: The Struggle to Win the Peace." *Asian Survey* **43**(1): 174–185.

Jarstad, A. K., and T. D. Sisk, Eds. (2008). *From War to Democracy: Dilemmas of Peacebuilding*. Cambridge, Cambridge University Press.

Jensen, E. G. (2003). "The Rule of Law and Judicial Reform: The Political Economy of Diverse Institutional Patterns and Reformers' Responses." In *Beyond Common Knowledge: Empirical Approaches to the Rule of Law*, ed. T. C. Heller and E. G. Jensen. Stanford, Stanford University Press: 336–381.

Jensen, E. G. (2008). "Justice and the Rule of Law." In *Building States to Build Peace*, ed. C. T. Call and V. Wyeth. Boulder, Lynne Rienner: 119–142.

Jensen, M. C., and W. H. Meckling (1976). "Theory of the Firm: Managerial Behavior, Agency Costs and Ownership Structure." *Journal of Financial Economics* 3(4): 305–360.

Johnson, R. (2011). *The Afghan Way of War: Culture and Pragmatism; A Critical Histroy*. London, Hurst.

Johnson, T. H. (2007). "The Taliban Insurgency and an Analysis of Shabnamah (Night Letters)." *Small Wars & Insurgencies* **18**(3): 317–344.

Johnson, T. H. (2013). "Taliban Adaptations and Innovations." *Small Wars & Insurgencies* **24**(1): 3–27.

Johnson, T. H., and M. C. DuPee (2012). "Analysing the New Taliban Code of Conduct (Layeha): An Assessment of Changing Perspectives and Strategies of the Afghan Taliban." *Central Asian Survey* **31**(1): 77–91.

Johnson, T. H., M. C. DuPee, M., and W. Shaaker (2017) *Taliban Narratives: The Use and Power of Stories in the Afghanistan Conflict*. Oxford, Oxford University Press.

Johnson, T. H. (2018) "The Myth of Afghan Electoral Democracy: The Irregularities of the 2014 Presidential Election," *Small Wars & Insurgencies* **29**(5–6): 1006–1039.

Johnson, T. H., and M. C. Mason (2007). "Understanding the Taliban and Insurgency in Afghanistan." *Orbis* **51**(1): 71–89.

Johnson, T. H., and M. C. Mason (2008). "No Sign until the Burst of Fire: Understanding the Pakistan-Afghanistan Frontier." *International Security* 32(4): 41–77.

Jolliffe, J. (1978). *East Timor: Nationalism and Colonialism*. St. Lucia, University of Queensland Press.

Jones, G. W. (2000). "East Timor: Education and Human Resource Development." In *Out of the Ashes: Destruction and Reconstruction of East Timor*, ed. J. J. Fox and D. Babo Soares. Canberra, Australian National University E-Press: 41–52.

Jones, L. (2010). "(Post-)Colonial State-Building and State Failure in East Timor: Bringing Social Conflict Back In." *Conflict, Security & Development* **10**(4): 547–575.

Jones, R. A. (1990). *The Soviet Concept of "Limited Sovereignty" from Lenin to Gorbachev: The Brezhnev Doctrine*. Houndsmills, Macmillan.

Jones, S. G., et al. (2005). *Establishing Law and Order after Conflict*. Santa Monica, RAND.
Jones, S. G. (2006). "Averting Failure in Afghanistan." *Survival* **48**(1): 111–128.
Jones, S. G. (2008). "The Rise of Afghanistan's Insurgency: State Failure and Jihad." *International Security* 32(4): 7–40.
Jones, S. G. (2010). *In the Graveyard of Empires: America's War in Afghanistan*. New York, W. W. Norton.
Jordan, L., and P. Van Tuijl, P. (2000). "Political Responsibility in Transnational NGO Advocacy." *World Development* **28**(12): 2051–2065.
Joshi, M. (2013). "Inclusive Institutions and Stability of Transition toward Democracy in Post-Civil War States." *Democratization* **20**(4): 743–770.
JSMP (2004). Justice Update: New Court Actor Training Program. Dili, JSMP.
JSMP (2009). The Crisis 2006: A Lesson for the Future. Dili, JSMP.
JSMP (2010). JSMP's Observations on Progress Achieved to Date and Challenges Facing the Legal System in Timor Leste. Dili, JSMP.
JSMP (2012). Overview of the Justice Sector 2012. Dili, JSMP.
JSMP (2013). Overview of the Justice Sector 2013. Dili, JSMP.
JSMP (2014). Overview of the Justice Sector in Timor-Leste 2013. Dili, JSMP.
JSMP (2017). Overview of the Justice Sector in Timor-Leste 2016. Dili, JSMP.
Justice Facility (2011). 2011–12 Work Plan and Resource Schedule. Dili, Justice Facility.
Kakar, H. K. (1979). *Government and Society in Afghanistan: The Reign of Amir 'Abd Al-Rahman Khan*. Austin, University of Texas Press.
Kakar, M. H. (1995). *Afghanistan: The Soviet Invasion and the Afghan Response, 1979–1982*. Berkeley, University of California Press.
Kakar, P. (2004). Tribal Law of Pashtunwali and Women's Legislative Authority. Afghan Legal History Project, Harvard Law School. Cambridge, Harvard Law School.
Kalyvas, S. N. (2006). *The Logic of Violence in Civil War*. Cambridge, Cambridge University Press.
Kamali, M. H. (1985). *Law in Afghanistan: A Study of the Constitutions, Matrimonial Law and the Judiciary*. Leiden, Brill.
Kamali, M. H. (2014). Afghanistan's Constitution Ten Years On: What Are the Issues? Kabul, Afghanistan Research and Evaluation Unit.
Kammen, D. (2015). "Timor-Leste." *The Contemporary Pacific* 27(2): 537–544.
Katzenstein, S. (2003). "Hybrid Tribunals: Searching for Justice in East Timor." *Harvard Human Rights Journal* **16**: 245–278.
Katzman, K. (2015). Afghanistan: Post-Taliban Governance, Security, and U.S. Policy. Washington, DC, Congressional Research Service.
Keane, C., and G. Diesen (2015). "Divided We Stand: The US Foreign Policy Bureaucracy and Nation-Building in Afghanistan." *International Peacekeeping* 22(3): 205–229.
Kelemen, R. D. (2020) "The European Union's Authoritarian Equilibrium." *Journal of European Public Policy* (27)3: 481–499.
Kent, L. (2012a). *The Dynamics of Transitional Justice: International Models and Local Realities in East Timor*. New York, Routledge.
Kent, L. (2012b). "Interrogating the 'Gap' between Law and Justice: East Timor's Serious Crimes Process." *Human Rights Quarterly* 34(4): 1021–1044.
Kent, L. (2014). "Beyond 'Pragmatism' Versus 'Principle': Ongoing Justice Debates in East Timor." In *Transitional Justice in the Asia Pacific*, ed. R. Jeffery and H. Joon Kim. Cambridge, Cambridge University Press: 157–194.
Khalilzad, Z. (1997). "Anarchy in Afghanistan." *Journal of International Affairs* **51**(1): 37.

Ki-Moon, B. (2009). Report of the Secretary-General on the UN Integrated Mission in Timor-Leste (for the Period from 21 January to 23 September 2009). New York, UN.

Kilcullen, D. J. (2011). Deiokes and the Taliban: Local Governance, Bottom-up State Formation and the Rule of Law in Counter-Insurgency. In *The Rule of Law in Afghanistan: Missing in Inaction*, ed. W. Mason. Cambridge, Cambridge University Press: 35–49.

Kingdom of Afghanistan (1923). The Constitution of Afghanistan. Kabul, Kingdom of Afghanistan.

Kingdom of Afghanistan (1931). Constitution of Afghanistan: Fundamental Principles of the Government of Afghanistan. Kabul, Kingdom of Afghanistan.

Kingdom of Afghanistan (1964). Constitution of Afghanistan. Kabul, Kingdom of Afghanistan.

Kingsbury, D. (2013). "The Constitution: Clarity without Convention." In *The Politics of Timor-Leste: Democratic Consolidation after Intervention*, ed. M. Leach and D. Kingsbury. Ithaca, Cornell Southeast Asia Program Publications: 69–84.

Kleinfeld, R. (2006). "Competing Definitions of the Rule of Law." In *Promoting the Rule of Law Abroad: The Search for Knowledge*, ed. T. Carothers. Washington, DC, Carnegie Endowment for International Peace: 31–73.

Kleinfeld, R. (2012). *Advancing the Rule of Law Abroad: Next Generation Reform*. Washington, DC, Carnegie Endowment for International Peace.

Knack, S. (2004). "Does Foreign Aid Promote Democracy?" *International Studies Quarterly* **48**(1): 251–266.

Kochanski, A. (2020). "The 'Local Turn' in Transitional Justice: Curb the Enthusiasm." *International Studies Review* **22**(1): 26–50.

Köhler, G., and N. Alcock (1976). "An Empirical Table of Structural Violence." *Journal of Peace Research* **13**(4): 343–356.

Kramer, L. D. (2001). "Foreword: We the Court." *Harvard Law Review* **115**(4): 5–169.

Krasner, S. D. (2004). "Sharing Sovereignty: New Institutions for Collapsed and Failing States." *International Security* **29**(2): 85–120.

Krasner, S. D., and T. Risse (2014). "External Actors, State-building, and Service Provision in Areas of Limited Statehood: Introduction." *Governance* **27**(4): 545–567.

Krent, H. J. (2001). "Conditioning the President's Conditional Pardon Power." *California Law Review* **89**(6): 1665–1720.

Kriger, N. (2003). *Guerrilla Veterans in Post-War Zimbabwe: Symbolic and Violent Politics, 1980–1987*. Cambridge, Cambridge University Press.

Kriger, N. (2006). "From Patriotic Memories to 'Patriotic History' in Zimbabwe, 1990–2005." *Third World Quarterly* **27**(6): 1151–1169.

Kyed, H. M. (2011). "Introduction to the Special Issue: Legal Pluralism and International Development Interventions." *The Journal of Legal Pluralism and Unofficial Law* **43**(63): 1–23.

Kyed, H. M., and L. Buur (2007). "Introduction: Traditional Authority and Democratization in Africa." In *State Recognition and Democratization in Sub-Saharan Africa: A New Dawn for Traditional Authorities*, ed. L. Buur and H. M. Kyed. New York, Palgrave Macmillan: 1–28.

Laffont, J. J., and D. Martimort (2002). *The Theoy of Incentives: The Principal Agent Problem*. Princeton, Princeton University Press.

Lake, D. A. (2008). "The State and International Relations." In *The Oxford Handbook of International Relations*, ed. C. Reus-Smit and D. Snidal Duncan. Oxford, Oxford University Press: 41–61.

Lambourne, W. (2009). "Transitional Justice and Peacebuilding after Mass Violence." *International Journal of Transitional Justice* 3(1): 28–48.
Lancaster, T. D. (1986). "Electoral Structures and Pork Barrel Politics." *International Political Science Review* 7(1): 67–81.
Larsen, I. H. (2010). *UNAMA in Afghanistan: Challenges and Opportunities in Peacemaking, State-Building and Coordination*. Olso, Norwegian Institute of International Affairs.
Larson, A. (2015). Political Parties in Afghanistan. Washington, DC, United States Institute for Peace.
Lawrence, J. P. (2018). "Afghan Anti-Corruption Program Is Corrupt, US Officials Say." November 9, *Stars and Stripes*. Retrieved September 20, 2020, from https://www.stripes.com/news/afghan-anti-corruption-program-is-corrupt-us-officials-say-1.555894.
Leach, M. (2006). "East Timorese History after Independence." *History Workshop Journal* 61(1): 222–237.
Ledwidge, F. (2009). "Justice and Counter-Insurgency in Afghanistan: A Missing Link." *The RUSI Journal* 154(1): 6–9.
Leeth, J., et al. (2012). Rule of Law Stabilization—Formal Sector Component Program Evaluation. Kabul, USAID.
Legal Adviser from the RDTL President's Office (2014). Interview with a Legal Adviser from the RDTL President's Office, Dili.
Levitsky, S., and L. Way (2010). *Competitive Authoritarianism*. Cambridge, Cambridge University Press.
Licht, A. N., et al. (2007). "Culture Rules: The Foundations of the Rule of Law and Other Norms of Governance." *Journal of Comparative Economics* 35(4): 659–688.
Lijphart, A., et al. (1986). "The Limited Vote and the Single Nontransferable Vote: Lessons from the Japanese and Spanish Examples." In *Electoral Laws and Their Political Consequences*, ed. B. Grofman and A. Lijphart. New York, Agathon Press: 154–169.
Linz, J. J. (1994). "Presidential or Parliamentary Democracy: Does It Make a Difference?" In *The Failure of Presidential Democracy: Comparative Perspectives, Volume 1*, ed. J. J. Linz and A. Valenzuela. Baltimore, The Johns Hopkins University Press: 3–87.
Lipset, S. M. (2000). "The Indispensability of Political Parties." *Journal of Democracy* 11(1): 48–55.
Local NGO Professional (2014). Interview with Local NGO Professional in Afghanistan. Kabul.
Lothe, E., and G. Peake (2010). "Addressing Symptoms but Not Causes: Stabilisation and Humanitarian Action in Timor-Leste." *Disasters* 34: S427–S443.
Loveman, M. (2005). "The Modern State and the Primitive Accumulation of Symbolic Power." *American Journal of Sociology* 110(6): 1651–1683.
Low, S. (2007). Evaluation of the "Providing Access to Justice: Legal Awareness at the Grassroots Level." Project Timor Leste for Avocats sans Frontières Brussels. Brussels, ASF Brussels.
Lundahl, M., and F. Sjöholm (2008). "The Oil Resources of Timor-Leste: Curse or Blessing?" *The Pacific Review* 21(1): 67–86.
Mac Ginty, R. (2008). "Indigenous Peace-Making Versus the Liberal Peace." *Cooperation and Conflict* 43(2): 139–163.
Mac Ginty, R. (2010). "Warlords and the Liberal Peace: State-Building in Afghanistan." *Conflict, Security & Development* 10(4): 577–598.
Mac Ginty, R. (2011). *International Peacebuilding and Local Resistance: Hybrid Forms of Peace*. London, Palgrave Macmillan.

Maia, V., et al. (2012). Trends of Local Governance in Timor-Leste: Suco Governance Performance Scale. Dili, Asia Foundation.

Mainwaring, S., and M. S. Shugart (1997). *Presidentialism and Democracy in Latin America*. Cambridge, Cambridge University Press.

Malejacq, R. (2019). *Warlord Survival: The Delusion of State-Building in Afghanistan*. Ithaca, Cornell University Press.

Maley, W. (2011). "Afghanistan in 2010." *Asian Survey* 51(1): 85–96.

Maley, W. (2018). "Institutional Design, Neopatrimonialism, and the Politics of Aid in Afghanistan." *Asian Survey* 58(6): 995–1015.

Maley, W., and F. H. Saikal (1992). *Political Order in Post-Communist Afghanistan*. Boulder, Lynne Rienner.

Malinowski, B. (1985) [1926]. *Crime and Custom in Savage Society*. Totowa, Helix Books.

Mampilly, Z. C. (2011). *Rebel Rulers: Insurgent Governance and Civilian Life during War*. Ithaca, Cornell University Press.

Manning, C. (2006). "Political Elites and Democratic State-Building Efforts in Bosnia and Iraq." *Democratization* 13(5): 724–738.

Massoud, M. F. (2013). *Law's Fragile State: Colonial, Authoritarian, and Humanitarian Legacies in Sudan*. Cambridge, Cambridge University Press.

Massoud M.F. (2014). "International Arbitration and Judicial Politics in Authoritarian States." *Law & Social Inquiry* 39(1):1–30.

MSD (2010). Justice Institutions Strengthening Program, 3rd Quarterly Report—FY2010. Dili, MSD.

MSD (2012). Justice Institutions Strengthening Program: Project Completetion Report. Dili, MSD.

Management Systems International (2020). Assistance for the Development of Afghan Legal Access and Transparency: Midterm Performance Evaluation. USAID, Washington, D.C.

Mann, M. (2012). *The Sources of Social Power, Volume 3: Global Empires and Revolution, 1890–1945*. Cambridge, Cambridge University Press.

Manning, C. (2006). "Political Elites and Democratic State-Building Efforts in Bosnia and Iraq." *Democratization* 13(5): 724–738.

Marlowe, A. (2010). "Shura to Fail? Why US Officials Taking Tea with Local Afghan Leaders Seem to Be Wasting Their Time." *The New Republic*, May 23. Washington, DC. Retrieved June 27, 2022, from https://newrepublic.com/article/74942/shura-fail.

Marriott, A. (2009). "Legal Professionals in Development: Timor-Leste's Legislative Experiment." *Conflict, Security & Development* 9(2): 239–263.

Marriott, A. (2012). "Justice Sector Dynamics in Timor-Leste: Institutions and Individuals." *Asian Politics & Policy* 4(1): 53–71.

Marsden, P. (1998). *The Taliban: War, Religion and the New Order in Afghanistan*. London, Zed Books.

Marten, K. (2006). "Warlordism in Comparative Perspective." *International Security* 31(3): 41–73.

Martins, M. (2011). "Rule of Law in Iraq and Afghanistan?" *Army Lawyer* 2011(11): 21–27.

Maru, V. (2006). "Between Law and Society: Paralegals and the Provision of Justice Services in Sierra Leone and Worldwide." *Yale Journal of International Law* 31: 427–476.

Marx, S. (2013). Timor-Leste Law and Justice Survey 2013. Dili, Asia Foundation.

Mashal, M., et al. (2015). "Afghans Forming Militias to Fight against Taliban." *New York Times*, December 21.

Mason, C. (2014). "Fraud and Folly in Afghanistan." *Foreign Policy*, September 23. Retrieved June 27, 2022, from https://foreignpolicy.com/2014/09/23/fraud-and-folly-in-afghanistan/.

Mason, W., Ed. (2011). *The Rule of Law in Afghanistan: Missing in Inaction*. Cambridge, Cambridge University Press.

May, C. (2014). *The Rule of Law: The Common Sense of Global Politics*. Cheltenham, Edward Elgar.

McAuliffe, P. (2011). "UN Peace-Building, Transitional Justice and the Rule of Law in East Timor: The Limits of Institutional Responses to Political Questions." *Netherlands International Law Review* **58**(01): 103–135.

McChrystal, S. (2009). Commander's Initial Assessment. Kabul, International Security Assistance Force.

McDevitt, A., and Adib, E. (2021). National Corruption Survey 2020: Afghans' Perceptions and Experiences of Corruption. Kabul, Integrity Watch Afghanistan.

McEvoy, K. (2007). "Beyond Legalism: Towards a Thicker Understanding of Transitional Justice." *Journal of Law and Society* **34**(4): 411–440.

Mcloughlin, C. (2015). "When Does Service Delivery Improve the Legitimacy of a Fragile or Conflict-Affected State?" *Governance* **28**(3): 341–356.

McWilliam, A. (2005). "Houses of Resistance in East Timor: Structuring Sociality in the New Nation." *Anthropological Forum* **15**(1): 27–44.

McWilliam, A., and A. Bexley (2008). "Performing Politics: The 2007 Parliamentary Elections in Timor Leste." *Asia Pacific Journal of Anthropology* **9**(1): 66–82.

Meitzner Yoder, L. S. (2007). "Hybridising Justice: State-Customary Interactions over Forest Crime and Punishment in Oecusse, East Timor." *Asia Pacific Journal of Anthropology* **8**(1): 43–57.

Mendeloff, D. (2004). "Truth-Seeking, Truth-Telling, and Postconflict Peacebuilding: Curb the Enthusiasm?" *International Studies Review* **6**(3): 355–380.

Menkhaus, K. (2007). "Governance without Government in Somalia: Spoilers, State Building, and the Politics of Coping." *International Security* **31**(3): 74–106.

Merry, S. E. (1988). "Legal Pluralism." *Law & Society Review* **22**(5): 869–896.

Merry, S. E. (1991). "Law and Colonialism (Book Review)." *Law and Society Review* **25**(4): 889–922.

Merryman, J. H., and R. Pérez-Perdomo (2007). *The Civil Law Tradition: An Introduction to the Legal Systems of Europe and Latin America*. Stanford, Stanford University Press.

Miakhel, S., and N. Coburn (2010). Many Shuras Do Not a Government Make: International Community Engagement with Local Councils in Afghanistan. Washington, DC, United States Institute for Peace.

Migdal, J. S. (1988). *Strong Societies and Weak States: State-Society Relations and State Capabilities in the Third World*. Princeton, Princeton University Press.

Migdal, J. S. (2001). *State in Society: Studying How States and Societies Transform and Constitute One Another*. Cambridge, Cambridge University Press.

Ministry of Justice (2009). Draft National Policy on Relations between the Formal Justice System and Dispute Resolution Councils. Kabul, Islamic Republic of Afghanistan.

Ministry of Justice (2010). Draft Law on Dispute Resolution Shuras and Jirgas. Kabul, Islamic Republic of Afghanistan.

Ministry of Justice (2012). Ministry of Justice Official Gazette 1075. Kabul, Islamic Republic of Afghanistan.

Ministry of Justice (2015). Draft Law on Conciliation in Civil Disputes. Kabul, Islamic Republic of Afghanistan.

Ministry of Justice (2021). "Minister of Justice: For the Provision of Informal Justice We Have No Choice but to Legalize the Jirgas (Councils)." Retrieved October 26, 2021 from, https://moj.gov.af/en/minister-justice-provision-informal-justice-we-have-no-choice-legalize-jirgas-councils.

Moghadam, V. M. (2002). "Patriarchy, the Taleban, and Politics of Public Space in Afghanistan." *Women's Studies International Forum* **25**(1): 19–31.

Moore, S. (2001). "The Indonesian Military's Last Years in East Timor: An Analysis of Its Secret Documents." *Indonesia* (72): 9–44. https://www.jstor.org/stable/3351480?seq=1.

Moore, E. P. (1993). "Gender, Power, and Legal Pluralism: Rajasthan, India." *American Ethnologist* **20**(3): 522–542.

Moore, S. F. (1986). *Social Facts and Fabrications: "Customary" Law on Kilimanjaro, 1880–1980*. Cambridge, Cambridge University Press.

Mosse, D. (2006). "Anti-Social Anthropology? Objectivity, Objection, and the Ethnography of Public Policy and Professional Communities." *Journal of the Royal Anthropological Institute* **12**(4): 935–956.

Moustafa, T. (2007). T*he Struggle for Constitutional Power: Law, Politics, and Economic Development in Egypt*. New York, Cambridge University Press.

Moustafa, T. (2014). "Law and Courts in Authoritarian Regimes." *Annual Review of Law and Social Science* **10**: 281–299.

Mukhopadhyay, D. (2014). *Warlords, Strongman Governors, and the State in Afghanistan*. Cambridge, Cambridge University Press.

Murtazashvili, J. B. (2016). *Informal Order and the State in Afghanistan*. Cambridge, Cambridge University Press.

Mutalib, H. (2000). "Illiberal Democracy and the Future of Opposition in Singapore." *Third World Quarterly* **21**(2): 313–342.

Nachbar, T. B. (2011). "The Use of Law in Counterinsurgency." *Military Law Review* **213**: 140–164.

NATO (2011). Media Backgrounder: NATO Rule of Law Field Support Mission. Brussels, NATO.

Nevins, J. (2005). *A Not-So-Distant Horror: Mass Violence in East Timor*. Ithaca, Cornell University Press.

Nevins, J. (2007). "Timor-Leste in 2006: The End of the Post-Independence Honeymoon." *Asian Survey* **47**(1): 162–167.

Newell, R. (1986). "The Prospects for State Building in Afghanistan." In *The State, Religion and Ethnic Politics: Afghanistan, Iran, and Pakistan*, ed. A. Banuazizi and M. Weiner. Syracuse, Syracuse University Press: 104–123.

NGO Legal Specialist (2014). Interview with NGO Legal Specialist in Afghanistan. Kabul.

Nina, D. (2000). "Dirty Harry Is Back: Vigilantism in South Africa—The (Re) Emergence of the 'Good' and 'Bad' Community." *African Security Review* **9**(1): 18–28.

Niner, S. (2007). "Martyrs, Heroes and Warriors: The Leadership of East Timor." In *East Timor: Beyond Independence*, ed. D. Kingsbury and M. Leach. Clayton, Monash University Press: 113–128.

Niner, S. et. al. (2022). "Women's Political Participation in Post-Conflict Settings: The Case of Timor-Leste." *Asian Studies Review* **46**(2): 293–311.

Nixon, R. (2008). "Integrating Indigenous Approaches into a 'New Subsistence State': The Case of Justice and Conflict Resolution in East Timor." Ph.D. diss. Darwin, Charles Darwin University.

Nixon, R. (2012). *Justice and Governance in East Timor: Indigenous Approaches and the 'New Subsistence State.'* London, Routledge.
Nojumi, N. (2002). *The Rise of the Taliban in Afghanistan: Mass Mobilization, Civil War, and the Future of the Region*. New York, Palgrave.
Nojumi, N., et al. (2004). Afghanistan's Systems of Justice: Formal, Traditional, and Customary. Medford, Feinstein International Famine Center, Tufts University.
Nolan-Haley, J. M. (1996). "Court Mediation and the Search for Justice through Law." *Washington University Law Quarterly* 74: 47–102.
Norris, P. (2004). *Electoral Engineering: Voting Rules and Political Behavior*. Cambridge, Cambridge University Press.
North, D. C. (1990). *Institutions, Institutional Change and Economic Performance*. Cambridge, Cambridge University Press.
North, D. C., J. J. Wallis, and B. R. Weingast (2009). *Violence and Social Orders: A Conceptual Framework for Interpreting Recorded Human History*. Cambridge, Cambridge University Press.
O'Donnell, G. A. (2004). "Why the Rule of Law Matters." *Journal of Democracy* 15(4): 32–46.
Oaks, D. H. (1970). "Studying the Exclusionary Rule in Search and Seizure." *The University of Chicago Law Review* 37(4): 665–757.
Obuchi, T. (1987). "Role of the Court in the Process of Informal Dispute Resolution in Japan: Traditional and Modern Aspects, with Special Emphasis on in-Court Compromise." *Law Japan* 20: 74–101.
OECD Development Assistance Committee (2011). *OECD Development Assistance Peer Reviews OECD Development Assistance Peer Reviews: Portugal 2010*. Paris, OECD.
Office of the Special Representative for Afghanistan and Pakistan (2010). Afghanistan and Pakistan Regional Stabilization Strategy. Washington, DC, U.S. State Department.
Organski, A. F., and J. Kugler (1981). *The War Ledger*. Chicago, University of Chicago Press.
Ospina, S., and T. Hohe (2001). *Traditional Power Structures and the Community Empowerment and Local Governance Project: Final Report*. Dili, World Bank.
Ottaway, M., and A. Lieven (2002). *Rebuilding Afghanistan: Fantasy Versus Reality*. Washington, DC, Carnegie Endowment for International Peace.
Palacio, A. (2006). Legal Empowerment for the Poor: An Action Agenda for the World Bank. Washington, DC, The World Bank.
Paris, R. (2004). *At War's End: Building Peace after Civil Conflict*. Cambridge, Cambridge University Press.
Paris, R. (2009). "Does Liberal Peacebuilding Have a Future?" In *New Perspectives on Liberal Peacebuilding*, ed. E. Newman, R. Paris, and O. P. Richmond. Tokyo, United Nations University Press: 97–111.
Paris, R. (2010). "Saving Liberal Peacebuilding." *Review of International Studies* 36(02): 337–365.
Paris, R. (2013). "Afghanistan: What Went Wrong?" *Perspectives on Politics* 11(02): 538–548.
Paris, R., and T. D. Sisk, Eds. (2009a). *The Dilemmas of Statebuilding: Confronting the Contradictions of Postwar Peace Operations*. London, Routledge.
Paris, R., and T. D. Sisk (2009b). "Introduction: Understanding the Contradictions of Postwar Statebuilding." In *The Dilemmas of Statebuilding: Confronting the Contradictions of Postwar Peace Operations*, ed. R. Paris and T. D. Sisk. London, Routledge: 1–20.

Peake, G., et al. (2012). AusAID Timor-Leste Justice Sector Support Facility: Independent Completion Report. Dili, AusAID.

Peceny, M., and Y. Bosin (2011). "Winning with Warlords in Afghanistan." *Small Wars & Insurgencies* 22(4): 603–618.

Peerenboom R. (2002a). *China's Long March toward Rule of Law*. Cambridge, Cambridge University Press.

Peerenboom, R. (2002b). "Let One Hundred Flowers Bloom, One Hundred Schools Contend: Debating Rule of Law in China." *Michigan Journal of International Law* 23: 471–544.

Peletz, M. G. (2002). *Islamic Modern: Religious Courts and Cultural Politics in Malaysia*. Princeton, Princeton University Press.

Pereira, M., and M. M. Lete Koten (2012). "Hybrid Governance: Dynamics of Democracy at the 'Suku' Level." *Local-Global: Identity, Security, Community* 11: 222–232.

Peters, E. A., and J. M. Ubink (2015). "Restorative and Flexible Customary Procedures and Their Gendered Impact: A Preliminary View on Namibia's Formalization of Traditional Courts." *The Journal of Legal Pluralism and Unofficial Law* 47(2): 291–311.

Peters, G. (2011). "The Afghan Insurgency and Organized Crime." In *The Rule of Law in Afghanistan: Missing in Inaction*, ed. W. Mason. Cambridge, Cambridge University Press: 99–122.

Pinto, C. (2001). "The Student Movement and the Independence Struggle in East Timor: An Interview." In *Bitter Flowers, Sweet Flowers: East Timor, Indonesia, and the World Community*, ed. R. Tanter, M. Selden, and S. R. Shalom. Lanham, Rowman & Littlefield: 31–41.

Pinto, C., and M. Jardine (1997). *East Timor's Unfinished Struggle: Inside the Timorese Resistance*. Boston, South End Press.

Pollock, F., and F. W. Maitland (1923). *The History of English Law Before the Time of Edward I*. Cambridge, Cambridge University Press.

Portuguese Institute for Development Support (2010). *Portuguese Development Cooperation 2005–2010*. Lisbon, Portuguese Institute for Development Support.

Posner, E. A., and A. Vermeule (2004). "Transitional Justice as Ordinary Justice." *Harvard Law Review* 117(3): 761–825.

Poullada, L. B. (1973). *Reform and Rebellion in Afghanistan, 1919–1929: King Amanullah's Failure to Modernize a Tribal Society*. Ithaca, Cornell University Press.

Přibáň, J. (2007). *Legal Symbolism: On Law, Time and European Identity*. Aldershot, Ashgate.

Provost, R. (2021). *Rebel Courts: The Administration of Justice by Armed Insurgents*. New York, Oxford University Press.

Rahbari, S. (2018). "From Normative Pluralism to a Unified Legal System in Afghanistan." *Asian Journal of Law and Society* 5(2): 289–314.

Rajah, J. (2012). *Authoritarian Rule of Law: Legislation, Discourse and Legitimacy in Singapore*. Cambridge, Cambridge University Press.

Ramos-Horta, J. (1987). *FUNU: The Unfinished Saga of East Timor*, Lawrenceville, The Red Sea Press.

Rashid, A. (1999). "The Taliban: Exporting Extremism." *Foreign Affairs* 78(6): 22–35.

Rashid, A. (2001). *Taliban: Militant Islam, Oil, and Fundamentalism in Central Asia*. New Haven, Yale University Press.

Rashid, A. (2008). *Descent into Chaos: The U.S. and the Disaster in Pakistan, Afghanistan, and Central Asia*. New York, Penguin.

Rawkins, P., and P. Hashmi (2014). Independent Mid-Term Evaluation: UNDP Afghanistan, Justice and Human Rights in Afghanistan, Phase Two, Final Report. Kabul, UNDP.

Rawkins, P., and M. A. Kamawi (2012). Independent End-of-Term Evaluation: UNDP Afghanistan Justice and Human Rights in Afghanistan; Project Period: 26 June 2009–30 June 2012; Final Report (Revised). Kabul, UNDP.

Raz, J. (2009). *The Authority of Law: Essays on Law and Morality*. Oxford, Oxford University Press.

RDTL (2002). Constitution of the Democratic Republic of Timor-Leste. Dili, RDTL.

RDTL (2004a). On Community Authorities. Dili, RDTL. *Decree Law No. 5/2004*.

RDTL (2004b). On the Election of Suco Chiefs and Suco Councils. Dili, RDTL. *Law No. 2/2004*.

RDTL (2004c). Recruitment and Training for the Professionals of the Judiciary and for the Office of the Public Defender. Dili, RDTL. *Decree Law No. 15/2004*.

RDTL (2006). Law on the Election of the National Parliament. Dili, RDTL. *Law 6/2006*.

RDTL (2008). Law on the Juridical Regime Governing Private Legal Professional and Lawyers Training. Dili, RDTL. *Law No. 11/2008*.

RDTL (2009a). Community Leadership and Their Election. Dili, RDTL. *Law 3/2009*.

RDTL (2009b). Penal Code. Dili, RDTL. *Decree-Law 19/2009*.

RDTL (2010a). Justice Sector Plan for Timor-Leste 2011–2030. Dili, RDTL.

RDTL (2010b). Law against Domestic Violence. Dili, RDTL. *Law 7/2010*.

RDTL (2011). State Budget 2012: Budget Overview Book 1. Dili, RDTL.

RDTL (2016). Dili, RDTL. *Law No. 9/2016*.

RDTL Court of Appeal (2009). Case No 2/Const/2009/Tr. Dili, RDTL Court of Appeal.

RDTL Ministry of State Administration and the Asia Foundation (2013). Reflections on Law No. 3/2009: Community Leadership and Their Election. Dili, Asia Foundation.

RDTL MIJ Official (2014). Interview with RDTL MOJ Official, Dili.

RDTL National Statistics Directorate and United Nations Population Fund (2011). Population and Housing Census of Timor-Leste, 2010, Volume 4: Suco Report. Dili, RDTL.

Redfern, A., et al. (2004). *Law and Practice of International Commercial Arbitration*. London, Sweet & Maxwell.

Reiger, C. (2006). "Hybrid Attempts at Accountability for Serious Crimes in Timor Leste." In *Transitional Justice in the Twenty-First Century: Beyond Truth Versus Justice*, ed. N. Roht-Arriaza and J. Mariezcurrena. Cambridge, Cambridge University Press: 143–170.

Reilly, B. (2001). *Democracy in Divided Societies: Electoral Engineering for Conflict Management*. Cambridge, Cambridge University Press.

Reilly, B. (2002). "Electoral Systems for Divided Societies." *Journal of Democracy* 13(2): 156–170.

Reiter, A. G., et al. (2013). "Transitional Justice and Civil War: Exploring New Pathways, Challenging Old Guideposts." *Transitional Justice Review* 1(1): 12.

Renders, M. (2012). *Consider Somaliland: State-Building with Traditional Leaders and Institutions*. Leiden, Brill.

Reno, W. (1998). *Warlord Politics and African States*. London, Lynne Rienner.

Republic of Indonesia (1987). East Timor after Integration. Jakarta, Department of Information, Republic of Indonesia.

Reynolds, A. (2002). *The Architecture of Democracy: Constitutional Design, Conflict Management, and Democracy*. Oxford, Oxford University Press.

Reynolds, A. (2006). "The Curious Case of Afghanistan." *Journal of Democracy* 17(2): 104–117.

Reynolds, A., and A. R. Wilder (2004). *Free, Fair or Flawed: Challenges for Legitimate Elections in Afghanistan*. Kabul, Afghanistan Research and Evaluation Unit.

Richmond, O. P. (2011). "De-Romanticising the Local, De-Mystifying the International: Hybridity in Timor Leste and the Solomon Islands." *The Pacific Review* 24(1): 115–136.

Richmond, O. P., and J. Franks (2008). "Liberal Peacebuilding in Timor Leste: The Emperor's New Clothes?" *International Peacekeeping* 15(2): 185–200.

Robinson, G. (2003). East Timor 1999 Crimes against Humanity. Dili, UN Office of the High Commissioner for Human Rights (OHCHR)/University of California Los Angeles.

Robinson, G. (2009). *"If You Leave Us Here, We Will Die": How Genocide Was Stopped in East Timor*. Princeton, Princeton University Press.

Robinson, G. (2011). "East Timor Ten Years On: Legacies of Violence." *The Journal of Asian Studies* 70(04): 1007–1021.

Röder, T. J. (2007). "Little Steps Forward: Some Remarks on the Rome Conference on the Rule of Law in Afghanistan." In *Max Planck Yearbook of United Nations Law*, ed. A. V. Bogdandy and R. Wolfrum. Leiden, Brill. 11: 307–312.

Rodriguez, F., and H. Anwari (2011). UNDP Justice and Human Rights in Afghanistan Project: Independent/External Mid-Term Evaluation Report. Kabul, UNDP.

Roll, K. (2014). "Encountering Resistance: Qualitative Insights from the Quantitative Sampling of Ex-Combatants in Timor-Leste." *PS: Political Science & Politics* 47(02): 485–489.

Roll, K., and Swenson, G. (2019). "Fieldwork after Conflict: Contextualising the Challenges of Access and Data Quality." *Disasters* 43(2): 240–60.

Rose-Ackerman, S. (1999). *Corruption and Government: Causes, Consequences, and Reform*. Cambridge, Cambridge University Press.

Rosenberg, G. N. (1992). "Judicial Independence and the Reality of Political Power." *The Review of Politics* 54(03): 369–398.

Rotberg, R. I. (2002). "The New Nature of Nation-State Failure." *Washington Quarterly* 25(3): 83–96.

Rothstein, B. (2011). *The Quality of Government: Corruption, Social Trust, and Inequality in International Perspective*. Chicago, University of Chicago Press.

Roy, O. (1990). *Islam and Resistance in Afghanistan*. Cambridge, Cambridge University Press.

Roy, O. (1994). *The Failure of Political Islam*. London, IB Tauris.

Rubin, B. R. (1995). *The Search for Peace in Aghanistan: From Buffer State to Failed State*. New Haven, Yale University Press.

Rubin, B. R. (2000). "The Political Economy of War and Peace in Afghanistan." *World Development* 28(10): 1789–1803.

Rubin, B. R. (2002). *The Fragmentation of Afghanistan: State Formation and Collapse in the International System*. New Haven, Yale University Press.

Rubin, B. R. (2008). "The Politics of Security in Postconflict Statebuilding." In *Building States to Build Peace*, ed. C. T. Call and V. Wyeth. Boulder, Lynne Rienner: 25–47.

Sahin, S. B. (2007). "Building the State in Timor-Leste." *Asian Survey* 47(2): 250–267.

Saikal, A. (2004). *Modern Afghanistan: A History of Struggle and Survival*. London, I.B. Tauris.

Saikal, A. (2012). "The UN and Afghanistan: Contentions in Democratization and Statebuilding." *International Peacekeeping* 19(2): 217–234.

Salim, A. (2015). "The Constitutionalization of Shari'a in Muslim Societies: Comparing Indonesia, Tunisia and Egypt." In *The Sociology of Shari'a: Case Studies from around the World*, ed. A. Possamai, J. T. Richardson, and B. S. Turner. Cham, Springer: 199–217.

Samuels, K. (2006). Rule of Law Reform in Post-Conflict Countries: Operational Initiatives and Lessons Learnt. Washington, DC, World Bank.

Sartori, G. (1962). "Constitutionalism: A Preliminary Discussion." *The American Political Science Review* **56**(4): 853–864.

Sartori, G. (1997). *Comparative Constitutional Engineering: An Inquiry into Structures, Incentives, and Outcomes*. New York, Palgrave Macmillian.

Scambary, J. (2006). A Survey of Gangs and Youth Groups in Timor-Leste: A Report Commissioned by Australia's Agency for International Development, AusAID. Canberra, AusAID.

Scambary, J. (2009). "Anatomy of a Conflict: The 2006–2007 Communal Violence in East Timor." *Conflict, Security & Development* **9**(2): 265–288.

Schmeidl, S. (2011). "Engaging Traditional Justice Mechnisms in Afghanistan: State-Building Opportunity of Dangerous Liason?" In *The Rule of Law in Afghanistan: Missing in Inaction*, ed. W. Mason. Cambridge, Cambridge University Press: 149–171.

Schueth, S., et al. (2014). Performance Evaluation of the Rule of Law Stablization-Informal Component Program. Kabul, USAID.

Scott, J. C. (1998). *Seeing like a State: How Certain Schemes to Improve the Human Condition Have Failed*. New Haven, Yale University Press.

Sedra, M. (2003). "New Beginning or Return to Arms? The Disarmament, Demobilization and Reintegration Process in Afghanistan." In *State Reconstruction and International Engagement in Afghanistan*. Bonn, ZEF-LSE Workshop.

Shahrani, M., Nazif (1984). "Introduction: Marxist 'Revolution'and Islamic Resistance in Afghanistan." In *Revolutions and Rebellions in Afghanistan: Anthropological Perspectives*, ed. M. N. Shahrani and R. L. Canfield. Berkeley, Institute of International Studies, University of California, Berkeley: 3–57.

Shahrani, M. N. (1986). "State Building and Social Fragmentation in Afghanistan: A Historical Perspective." In *The State, Religion, and Ethnic Politics: Afghanistan, Iran, and Pakistan*, ed. A. Banuazizi and M. Weiner. Syracuse, Syracuse University Press: 23–74.

Shahrani, N. M. (2015). "The Impact of the 2014 U.S.-NATO Withdrawal on the Internal Politics of Afghanistan: Karzai-Style Thugocracy or Taliban Theocracy?" *Asian Survey* **55**(2): 273–298.

Shavell, S. (1995). "Alternative Dispute Resolution: An Economic Analysis." *The Journal of Legal Studies* **24**(1): 1–28.

Sherman, J. (2008). "Afghanistan: Nationally Led Statebuilding." In *Building States to Build Peace*, ed. C. Call and V. Wyeth. Boulder, Lynne Rienner: 303–334.

Sidel, M. (2008). *Law and Society in Vietnam: The Transition from Socialism in Comparative Perspective*. New York, Cambridge University Press.

SIGAR (2009). Actions Needed for a More Strategic Approach to U.S. Judicial Security Assistance. Washington, DC, SIGAR.

SIGAR (2011). April 2011: Quarterly Report to the United States Congress. Washington, DC, SIGAR.

SIGAR (2013a). July 2013: Quarterly Report to the United States Congress. Washington, DC, SIGAR.

SIGAR (2013b). October 2013: Quarterly Report to the United States Congress. Washington, DC, SIGAR.

SIGAR (2014a). SIGAR 14-26 Audit Report: Support for Afghanistan's Justice Sector: State Department Programs Need Better Management and Stronger Oversight. Washington, DC, SIGAR.

SIGAR (2014b). SIGAR 15-22 Financial Audit: Department of State's Afghanistan Justice Sector Support Program: Audit of Costs Incurred by Pacific Architects and Engineers, Inc. Washington, DC, SIGAR.

SIGAR (2014c). SIGAR-14-27-Sp: USAID Assistance to Afghanistan Reconstruction: $13.3 Billion Obligated between 2002 and 2013. Washington, DC, SIGAR.

SIGAR (2015a). January 2015: Quarterly Report to the United States Congress. Washington, DC, SIGAR.

SIGAR (2015b). SIGAR 15-68 Audit Report: Rule of Law in Afghanistan: U.S. Agencies Lack a Strategy and Cannot Fully Determine the Effectiveness of Programs Costing More than $1 Billion. Washington, DC, SIGAR.

SIGAR (2018). Stabilization: Lessons from the U.S. Experience in Afghanistan. Washington, DC, SIGAR.

SIGAR (2021a). What We Need to Learn: Lessons *from Twenty Years of Afghanistan Reconstruction*. Washington, DC, SIGAR.

SIGAR (2021b). The Risk of Doing the Wrong Thing Perfectly: Monitoring and Evaluation of Reconstruction Contracting in Afghanistan. Washington, DC, SIGAR.

Singh, D. (2015). "Explaining Varieties of Corruption in the Afghan Justice Sector." *Journal of Intervention and Statebuilding* 9(2): 231–255.

Sinno, A. H. (2008). *Organizations at War in Afghanistan and Beyond*. Ithaca, Cornell University Press.

Sisk, T. D. (2009). "Pathways of the Political: Electoral Processes after Civil War." In *The Dilemmas of Statebuilding: Confronting the Contradictions of Postwar Peace Operations*, ed. R. Paris and T. D. Sisk. London, Routledge: 196–223.

Skarbek, D. (2011). "Governance and Prison Gangs." *American Political Science Review* 105(04): 702–716.

Slater, D. (2012). "Strong-State Democratization in Malaysia and Singapore." *Journal of Democracy* 23(2): 19–33.

Smith, C. E. (1997). "Law and Symbolism." *Detior College of Law Review* 1997(3): 935–1281.

Smith, M. G., and M. Dee (2003). *Peacekeeping in East Timor: The Path to Independence*. London, Lynne Rienner.

Smith, S. S. (2020). *Service Delivery in Taliban-Influenced Areas of Afghanistan*. Washington, DC, United States Institute of Peace.

Soares, A. D. J. (2013). "Combating Corruption: Avoiding 'Institutional Ritualism.'" In *The Politics of Timor-Leste: Democratic Consolidation after Intervention*, ed. M. Leach and D. Kingsbury. Ithaca, Southeast Asia Program Publications, Cornell University: 85–97.

Sriram, C. L. (2007). "Justice as Peace? Liberal Peacebuilding and Strategies of Transitional Justice." *Global Society* 21(4): 579–591.

Stanger, A. (2009). *One Nation under Contract: The Outsourcing of American Power and the Future of Foreign Policy*. New Haven, Yale University Press.

Staniland, P. (2012). "States, Insurgents, and Wartime Political Orders." *Perspectives on Politics* 10(02): 243–264.

Stepan, A., and C. Skach (1993). "Constitutional Frameworks and Democratic Consolidation: Parliamentarianism Versus Presidentialism." *World Politics* 46(1): 1–22.

Sternlight, J. R. (2005). "Creeping Mandatory Arbitration: Is It Just?" *Stanford Law Review* 57(5): 1631–1675.

Stipanowich, T. J. (2004). "ADR and the 'Vanishing Trial': The Growth and Impact of 'Alternative Dispute Resolution.'" *Journal of Empirical Legal Studies* 1(3): 843–912.
Stone, D. A. (1989). "Causal Stories and the Formation of Policy Agendas." *Political Science Quarterly* **104**(2): 281–300.
Strohmeyer, H. (2000). "Building a New Judiciary for East Timor: Challenges of a Fledgling Nation." *Criminal Law Forum* 11(3): 259–285.
Strohmeyer, H. (2001a). "Collapse and Reconstruction of a Judicial System: The United Nations Missions in Kosovo and East Timor." *The American Journal of International Law* **95**(1): 46–63.
Strohmeyer, H. (2001b). "Policing the Peace: Post-Conflict Judicial System Reconstruction in East Timor." *University of New South Wales Law Journal* **24**: 171–182.
Stromseth, J., D. Wippman, and R. Brooks (2006). *Can Might Make Rights? Building the Rule of Law after Military Interventions*. Cambridge, Cambridge University Press.
Suhrke, A. (2008). "Democratizing a Dependent State: The Case of Afghanistan." *Democratization* 15(3): 630–648.
Suhrke, A. (2011). *When More Is Less: The International Project in Afghanistan*. London, Hurst.
Suhrke, A., and K. Borchgrevink (2009a). "Afghanistan: Justice Sector Reform." In *New Perspectives on Liberal Peacebuilding*, ed. E. Newman, R. Paris, and O. P. Richmond. Tokyo, United Nations Press: 178–200.
Suhrke, A., and K. Borchgrevink (2009b). "Negotiating Justice Sector Reform in Afghanistan." *Crime, Law and Social Change* 51(2): 211–230.
Swenson, G. (2017). "Why U.S. Efforts to Promote the Rule of Law in Afghanistan Failed." *International Security* 42(1): 114–51.
Swenson, G. (2018a). "Legal Pluralism in Theory and Practice." *International Studies Review* **20**(3): 438–62.
Swenson, G. (2018b). "The Promise and Peril of Paralegal Aid." *World Development* **106**: 51–63.
Swenson, G., and E. Sugerman (2011). "Building the Rule of Law in Afghanistan: The Importance of Legal Education." *Hague Journal on the Rule of Law* 3(01): 130–146.
Swenson, G., and J. Kniess (2021). "International Assistance after Conflict: Health, Transitional justice and Opportunity costs." *Third World Quarterly* **42**(8): 1696–1714,
Swenson, G., and K. Roll (2020). "Theorizing Risk and Research: Methodological Constraints and Their Consequences." *PS: Political Science & Politics* **53**(2): 286–291.
Tamanaha, B. Z. (2004). *On the Rule of Law: History, Politics, Theory*. Cambridge, Cambridge University Press.
Tamanaha, B. Z. (2007). A Concise Guide to the Rule of Law. St. John's Legal Studies Research Paper No. 07-0082..
Tamanaha, B. Z. (2008). "Understanding Legal Pluralism: Past to Present, Local to Global." *Sydney Law Review* **30**: 375–411.
Tamanaha, B. Z. (2011). "The Rule of Law and Legal Pluralism in Development." *Hague Journal on the Rule of Law* 3(01): 1–17.
Tamanaha, B. Z., et al., Eds. (2012). *Legal Pluralism and Development: Scholars and Practitioners in Dialogue*. Cambridge, Cambridge University Press.
Tamanaha, B. Z. (2021). *Legal Pluralism Explained: History, Theory, Consequences*. New York, Oxford University Press.
Tansey, O. (2007). "Process Tracing and Elite Interviewing: A Case for Non-Probability Sampling." *PS: Political Science & Politics* **40**(04): 765–772.

Tansey, O. (2009). *Regime-Building: Democratization and International Administration: Democratization and International Administration*. Oxford, Oxford University Press.

Tansey, O. (2014). "Evaluating the Legacies of State-Building: Success, Failure, and the Role of Responsibility." *International Studies Quarterly* 58(1): 174–186.

Tarzi, A. (2012). "Islam and Constitutionalism in Afghanistan." *Journal of Persianate Studies* 5(2): 205–243.

Tarzi, S. M. (1993). "Afghanistan in 1992: A Hobbesian State of Nature." *Asian Survey* 33(2): 165–174.

Taylor, J. G. (1991). *Indonesia's Forgotten War: The Hidden History of East Timor*. London, Zed Books.

Taylor, J. G. (1999). *East Timor: The Price of Freedom*. London, Zed Books.

Taylor-Leech, K. (2008). "Language and Identity in East Timor." *Language Problems & Language Planning* 32(2): 153–180.

Taylor-Leech, K. (2013). "Finding Space for Non-Dominant Languages in Education: Language Policy and Medium of Instruction in Timor-Leste 2000–2012." *Current Issues in Language Planning* 14(1): 109–126.

Tetra Tech DPK (2012). Afghanistan Rule of Law Stabilization Program (Formal Component): Performance Monitoring Plan July 2012 to January 2014. Kabul, USAID.

Tetra Tech DPK (2014). Rule of Law Stabilization (Formal Component): Final Report. Kabul, USAID.

Thoms, O. N. T., et al. (2010) "State-Level Effects of Transitional Justice: What Do We Know?" *International Journal of Transitional Justice*, 4(3): 329–354.

Tilly, C. (1992). *Coercion, Capital, and European States, AD 990–1992*. Oxford, Blackwell.

Tilman, M. (2012). "Customary Social Order and Authority in the Contemporary East Timorese Village: Persistence and Transformation." *Local-Global: Identity, Security, Community* 11: 192–205.

Tobe, R., et al., Eds. (2013). *Architecture and Justice: Judicial Meanings in the Public Realm*. Farnham, Ashgate.

Tondini, M. (2007). "Rebuilding the System of Justice in Afghanistan: A Preliminary Assessment." *Journal of Intervention and Statebuilding* 1(3): 333–354.

Tondini, M. (2010). *Statebuilding and Justice Reform: Post-Conflict Reconstruction in Afghanistan*. London, Routledge.

Transitional Islamic State of Afghanistan (2003). Political Parties Law. Kabul, Transitional Islamic State of Afghanistan.

Trefon, T. (2009). "Public Service Provision in a Failed State: Looking beyond Predation in the Democratic Republic of Congo." *Review of African Political Economy* 36(119): 9–21.

Tricia, D. O., et al. (2010). "The Justice Balance: When Transitional Justice Improves Human Rights and Democracy." *Human Rights Quarterly* 32(4): 980–1007.

Trindade, J., and B. Castro (2007). Rethinking Timorese Identity as a Peacebuilding Strategy: The Lorosa'e–Loromonu Conflict from a Traditional Perspective. Dili, GTZ.

Tsebelis, G. (2002). *Veto Players: How Political Institutions Work*. Princeton, Princeton University Press.

Tyler, T. R. (2006). *Why People Obey the Law*. Princeton, Princeton University Press.

U.N. Independent Special Commission of Inquiry for Timor-Leste (2006). Report of the United Nations Independent Special Commission of Inquiry for Timor-Leste. Geneva, United Nations.

U.N. Commission on Legal Empowerment of the Poor (2008). *Making the Law Work for Everyone*. New York, United Nations.

UNAMA (2015a). "Human Rights." Retrieved August 4, 2015, from http://unama.unmissions.org/Default.aspx?tabid=12285&language=en-US.
UNAMA (2015b). "Rule of Law." Retrieved August 10, 2015, from http://unama.unmissions.org/Default.aspx?tabid=12287&language=en-US.
UNDP (2002). Timor-Leste Correctional Service: Setting the Course. Dili, UNDP.
UNDP (2004). Afghanistan: National Human Development Report 2004. Kabul, UNDP.
UNDP (2005). Rebuilding the Justice Sector of Afghanistan: Quarterly Report; First Quarter (January to March) 2005. Kabul, UNDP.
UNDP (2008). United Nations Development Programme in Timor-Leste (2008–2013). Dili, UNDP.
UNDP (2009a). Project Document: Justice and Human Rights in Afghanistan Project (Jhra): 26 June 2009–30 June 2012 (36 Months). Kabul, UNDP.
UNDP (2009b). Strengthening the Justice System of Afghanistan: First Quarter 2009 Project Report. Kabul, UNDP.
UNDP (2011a). "Strengthening the Justice System in Timor-Leste." In *Human Rights Education in Asia-Pacfic: Volume Two*, ed. J. Plantilla. Osaka, HURIGHTS Osaka: 107–113.
UNDP (2011b). Strengthening the Rule of Law in Crisis-Affected and Fragile Situations: Global Programme Annual Report 2011. New York, UNDP.
UNDP (2012). Justice and Human Rights in Afghanistan Project: Final Report. Kabul, UNDP.
UNDP (2013a). Strengthening the Justice System in Timor-Leste (UNDP Project No. 00014955): Second Quarterly Progess Report. Dili, UNDP.
UNDP (2013b). Strengthening the Justice System in Timor-Leste (UNDP Project No. 00014955): Justice System Programme, 2012 Annual Report. Dili, UNDP.
UNDP (2013c). Strengthening the Rule of Law in Crisis-Affected and Fragile Situations: Global Programme Annual Report 2013. New York, UNDP.
UNDP et al. (2012). Informal Justice Systems: Charting a Course for Human Rights-Based Engagement. New York, UNDP, UN Women, UNICEF.
UNDP Official (2014). Interview with UNDP Official. Dili.
U.N. General Assembly (December 12, 1975). Question of Timor. *3485 (XXX)*. U.N.G. Assembly. New York, UN General Assembly.
U.N. Independent Comprehensive Needs Assessment Team (2009). The Justice System of Timor-Leste: An Independent Comprehensive Assessment. Dili, United Nations.
UNMIT (2008). Report on Human Rights Developments in Timor-Leste: The Security Sector and Access to Justice 1 September 2007–30 June 2008. Dili, UNMIT.
UNMIT (2011). Index of Laws of Timor-Leste with Internet Links: 20 May 2002–1 March 2011. Dili, UNMIT.
UNMIT (2012). "Momentum: Timor-Leste Forges a Vibrant Future." Retrieved March 5, 2013, from http://www.momentum.tl/en/index.html.
UNMIT and UNDP (2012). Compendium of the 2012 Elections in Timor-Leste: As of 21 June 2012. Dili, UNMIT/UNDP.
UNSC (1975). Resolution 384. New York, United Nations.
UNSC (1999). Resolution 1272. New York, United Nations.
UNSC (2002a). Report of the Secretary-General on the UN Mission of Support in East Timor. New York, UN.
UNSC (2002b). Resolution 1401. New York, United Nations.
UNSC (2002c). Resolution 1410. New York, United Nations.

UNSC (2005). Resolution 1599. New York, United Nations.

UNSC (2006). Resolution 1704. New York, United Nations.

UNSC (2011). Resolution 1969. New York, United Nations.

UNTAET (1999). On the Establishment of a National Consultative Council. *1999/2*. Dili, UNTAET.

UNTAET (2000a). On the Establishment of a National Council. *UNTAET/REG/2000/24*. Dili, UNTAET.

UNTAET (2000b). On the Establishment of Panels with Exclusive Jurisdiction over Serious Criminal Offences. *UNTAET/REG/2000/15*. Dili, UNTAET.

UNTAET (2000c). On the Establishment of the Cabinet of the Transitional Government in East Timor. *UNTAET/REG/2000/23*. Dili, UNTAET.

UNTAET (2000d). On the Organization of Courts in East Timor. *UNTAET/REG/2000/11*. Dili, UNTAET.

UNTAET (2000e). On the Organization of the Public Prosecutor Service in East Timor. *UNTAET/REG/2000/15*. Dili, UNTAET.

UNTAET (2001). On the Establishment of a Commission for Reception, Truth and Reconciliation in East Timor. *UNTAET/REG/2001/10*. Dili, UNTAET.

U.S. Department of Defense and U.S. Department of State (2009). United States Government Integrated Civilian-Military Campaign Plan for Support to Afghanistan. Kabul, U.S. Department of Defense and Department of State.

U.S. Government Accountability Office (2013). Afghanistan: Key Oversight Issues. Washington, DC, U.S. Government Accountability Office.

U.S. Mission Afghanistan (2010). U.S. Foreign Assistance for Afghanistan: Post Performance Management Plan 2011–2015. Kabul, U.S. Mission Afghanistan.

U.S. Senate Foreign Relations Committee (2011). Evaluating U.S. Foreign Assistance to Afghanistan: A Majority Staff Report. Washington, DC, U.S. Government Printing Office.

U.S. State Department (2015a). Afghanistan 2014 Human Rights Report. Washington, DC, U.S. State Department.

U.S. State Department (2015b). "INL: Afghanistan Program Overview." Retrieved July 9, 2015, from http://go.usa.gov/3y2Sk.

U.S. State Department Office of the Inspector General (2008). Report of Inspection: Rule-of-Law Programs in Afghanistan. Washington, DC, U.S. State Department.

USAID (2005). USAID/Afghanistan Strategic Plan. Washington, DC, USAID.

USAID (2015). Request for Proposals: Adalat Program. Washington, DC, USAID.

USAID Afghanistan (2018) Country Development Cooperation Strategy: FY2019–2023. Kabul, USAID.

Uslaner, E. M. (2008). *Corruption, Inequality, and the Rule of Law: The Bulging Pocket Makes the Easy Life*. Cambridge, Cambridge University Press.

Valuer, I. (1971). "The Roman Catholic Church: A Transnational Actor." *International Organization* **25**(03): 479–502.

Vinjamuri, L., and J. Snyder (2015). "Law and Politics in Transitional Justice", *Annual Review of Political Science* **18**: 303–327.

Volkov, V. (2000). "The Political Economy of Protection Rackets in the Past and the Present." *Social Research* **67**(3): 709–744.

Von Benda-Beckmann, F., and K. Von Benda-Beckmann (2006). "The Dynamics of Change and Continuity in Plural Legal Orders." *Journal of Legal Pluralism & Unofficial Law* **53–54**: 1–44.

von Benda-Beckmann, F., K. von Benda-Beckmann, and A. Griffiths, Eds. (2009). *The Power of Law in a Transnational World: Anthropological Enquiries*. New York, Berghahn Books.

Vyas, Y. (1992). "The Independence of the Judiciary: A Third World Perspective." *Third World Legal Studies* (11): 127–177.

Waldorf, L. (2006). "Mass Justice for Mass Atrocity: Rethinking Local Justice as Transitional Justice." *Temple Law Review* 79: 1–87.

Wardak, A. (2003). Jirga–A Traditional Mechanism of Conflict Resolution in Afghanistan. Glamorgan, University of Glamorgan. Retrieved June 27, 2022, from https://www.researchgate.net/profile/Ali-Wardak/publication/254937578_Jirga_-_A_Traditional_Mechanism_of_Conflict_Resolution_in_Afghanistan/links/574c0d1908ae1e99d0e4e551/Jirga-A-Traditional-Mechanism-of-Conflict-Resolution-in-Afghanistan.pdf.

Wardak, A. (2004). "Building a Post-War Justice System in Afghanistan." *Crime, Law and Social Change* 41(4): 319–341.

Wardak, A. (2011). "State and Non-State Justice Systems in Afghanistan: The Need for Synergy." *University of Pennsylvania Journal of International* Law 32(5): 1305–1324.

Wardak, A., and J. Braithwaite (2013). "Crime and War in Afghanistan: Part II: A Jeffersonian Alternative?" *British Journal of Criminology* 53(2): 197–214.

Wassel, T. (2014). Institutionalising Community Policing in Timor-Leste: Police Development in Asia's Youngest Country. London, Overseas Development Institute/Asia Foundation.

Weber, M. (1978). *Economy and Society: An Outline of Interpretive Sociology*. Berkeley, University of California Press.

Webster, D. (2003). "Non-State Diplomacy: East Timor 1975–99." *Portuguese Studies Review* 11(1): 1–28.

Weigand, F. (2017). "Afghanistan's Taliban—Legitimate Jihadists or Coercive Extremists?" *Journal of Intervention and Statebuilding* 11(3): 359–381.

Weingast, B. R. (1997). "The Political Foundations of Democracy and the Rule of Law." *American Political Science Review* 91(2): 245–263.

Weinstein, J. M. (2005). Autonomous Recovery and International Intervention in Comparative Perspective. Working Paper 57. Washington, DC, Center for Global Development.

West, R. (2003). *Re-Imagining Justice: Progressive Interpretations of Formal Equality, Rights, and the Rule of Law*. Farnham, Ashgate.

Weiner, J. F. (2006). "Eliciting Customary Law." *Asia Pacific Journal of Anthropology* 7(1): 15–25.

West, R. A. (2007). "Lawyers, Guns and Money: Justice and Security Reform in East Timor." In *Constructing Justice and Security after War*, ed. C. T. Call. Washington, DC, U.S. Institute for Peace: 313–350.

Whiting, S. H. (2017). "Authoritarian 'Rule of Law' and Regime Legitimacy." *Comparative Political Studies* 50(14): 1907–1940.

Wickham-Crowley, T. P. (1992). *Guerrillas and Revolution in Latin America: A Comparative Study of Insurgents and Regimes since 1956*. Princeton, Princeton University Press.

Wigglesworth, A. (2013). "Community Leadership and Gender Equality: Experiences of Representation in Local Governance in Timor-Leste." *Asian Politics & Policy* 5(4): 567–584.

Wilder, A. R. (2005). *A House Divided? Analysing the 2005 Afghan Elections*. Kabul, Afghanistan Research and Evaluation Unit.

Wilson, R. A. (2000). "Reconciliation and Revenge in Post-Apartheid South Africa." *Current Anthropology* 41(1): 75–98.

Wimpelmann, T. (2013). "Nexuses of Knowledge and Power in Afghanistan: The Rise and Fall of the Informal Justice Assemblage." *Central Asian Survey* 32(3): 406–422.

Wojkowska, E. (2006). Doing Justice: How Informal Systems Can Contribute. New York, UNDP.

Wolfowitz, P. (2002). Building a Better World: One Path from Crisis to Opportunity. Washington, DC.

Woodman, G. R. (1998). "Ideological Combat and Social Observation-Recent Debate About Legal Pluralism." *Journal of Legal Pluralism & Unofficial Law* 30(42): 21–59.

Worden, S. (2010). "Afghanistan: An Election Gone Awry." *Journal of Democracy* 21(3): 11–25.

World Bank (2003). Timor-Leste Poverty Assessment: Poverty in a New Nation: Analysis for Action. Washington, DC, World Bank. **Volume I: Main Report**.

World Bank (2006). Strengthening the Institutions of Goverance in Timor-Leste. Dili, World Bank.

World Bank (2011). *World Development Report 2011: Conflict, Security, and Development*. Washington, DC, World Bank.

Wyler, L. S., and K. Katzman (2010). Afghanistan: U.S. Rule of Law and Justice Sector Assistance. Washington, DC, Congressional Research Service.

Yin, R. K. (2009). *Case Study Research: Design and Methods*. London, Sage.

Zaum, D. (2007). *The Sovereignty Paradox: The Norms and Politics of International Statebuilding*. Oxford, Oxford University Press.

Zips, W. (2005). "'Global Fire': Repatriation and Reparations from a Rastafari (Re) Migrant's Perspective." In *Mobile People, Mobile Law: Expanding Legal Relations in a Contracting World*, ed. F. von Benda-Beckman, K. von Benda Beckman, and A. Griffiths. Aldershot, Ashgate: 81–102.

Zürcher, C. (2017). "What Do We (Not) Know about Development Aid and Violence? A Systematic Review." *World Development* 98:506–522.

Index

For the benefit of digital users, indexed terms that span two pages (e.g., 52–53) may, on occasion, appear on only one of those pages.

Tables are indicated by *t* following the page number